Hospital Stay
Health Care Made Simple

Craig B. Garner, Esq.
Thomas A. Gionis, M.D.
Galal S. Gough, M.D.

Published by To Sustain, LLC

Hospital Stay: Health Care Made Simple
First Published in 2010 by To Sustain, LLC
Copyright © 2010 Craig Boyd Garner
All rights reserved

Designed by To Sustain, LLC

Without limiting the rights under copyright reserved above, no part of this publication may be reproduced, stored, or introduced into a retrieval system, or transmitted, in any form, or by any means (electronic, mechanical, photocopying, recording, or otherwise), without the prior permission of both copyright owner and publisher.

The scanning, uploading, and distribution of this book via the Internet or via any other means without the permission of the copyright owner is illegal and punishable by law. Please purchase only authorized electronic editions and do not participate in or encourage electronic piracy of copyrightable materials.

Requests for permission should be directed to info@tosustain.com, or mailed to the copyright owner at To Sustain, LLC, 578 Washington Boulevard, #430, Marina del Rey, California 90292

ISBN: 978-0-557-61314-4

Library of Congress Cataloging-in-Publication Data Pending

Printed in the United States of America

To Sustain, LLC, Hospital Stay, colophons, and logos are registered trademarks of To Sustain, LLC

First Edition

In memory of my father, Gerald J. Garner, who made this book possible in the most unusual way.

TABLE OF CONTENTS

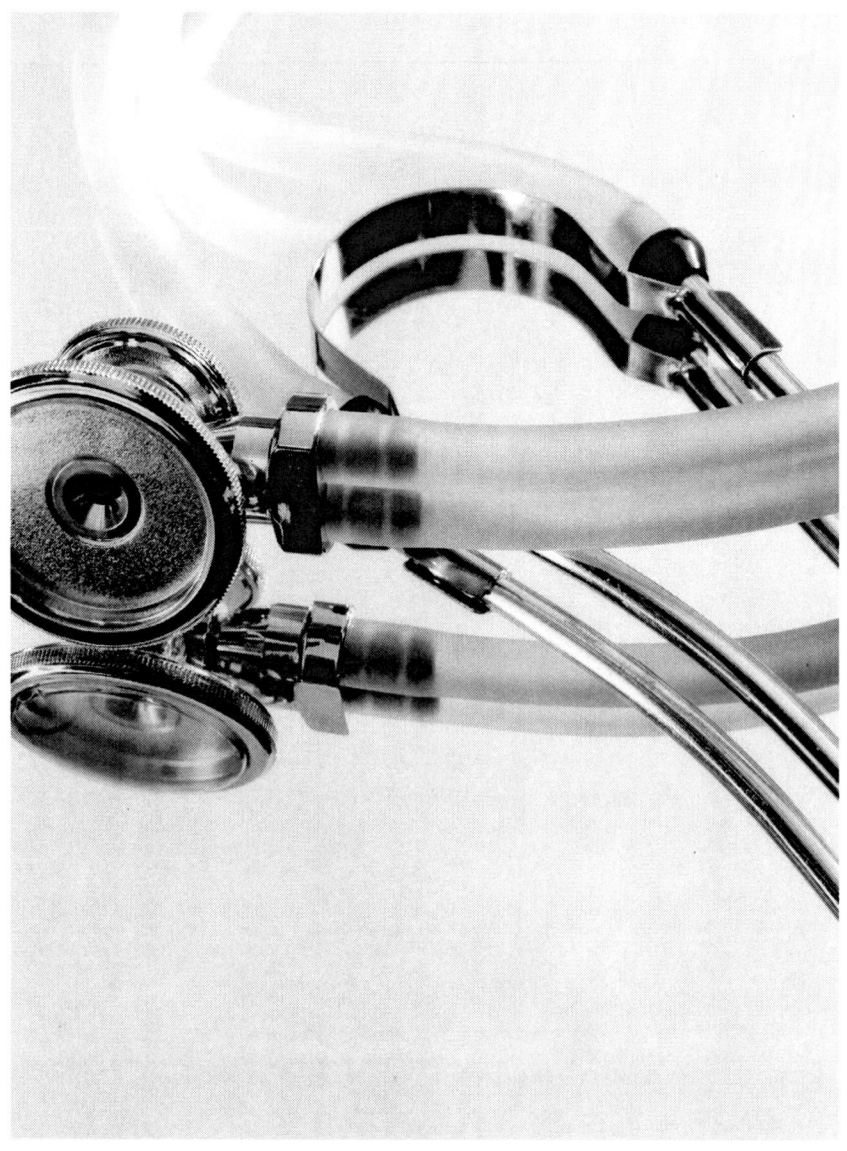

"Simplicity is the final achievement. After one has played a vast quantity of notes and more notes, it is simplicity that emerges as the crowning reward of art."
– Frédéric François Chopin, Polish composer and pianist

TABLE OF CONTENTS

Introduction	iii
The People in Your Hospital	2
The Emergency Room	12
Intravenous Therapy	20
Vital Signs	30
Radiology and Imaging	36
The Art of Diagnostic Medicine	50
Surgery Made Simple	62
Anesthesia	78
Hyperbaric Oxygen Therapy	86
The Language of Blood	92
The Hospital Menu	104
Building a Balanced Hospital	110
About Aspirin	118
Understanding Your Hospital Bill	128
Leaving the Hospital	138
Welcome to the Morgue	152
Complementary and Alternative Medicine	162

INTRODUCTION

A hospital stay in the United States can be a great equalizer. It mixes disease and diagnosis, joy and despair, and physical and mental tribulations, all of which are encompassed by matters of life and death. Regardless of class, education, or ethnicity, however, few people understand the science and service behind the delivery of modern medicine, let alone the billing and insurance systems that usually follow. As health care reform takes center stage across the nation, hospital stays continue to be full of uncertainty and trepidation.

For the past eight years, I have had a front row seat to what public opinion has perceived to be the decline of Western medicine, particularly as it applies to the health care system in the United States. As the successor to hospital CEO after my father's unexpected death, I transitioned from health care lawyer to the head of a small community hospital in Los Angeles County, California.

During this time, I have come to understand a great deal about the current state of health care in this country. One of the biggest surprises I have found is that health care and hospitals are not as complicated as most people believe. My inspiration to write this book came from Aleksandr Solzhenitsyn's *Cancer Ward*, a novel published in 1968 and set in a Soviet hospital circa 1955:

"Unforeseen and unprepared for, the disease had come upon him, a happy man with few cares, like a gale in the space of two weeks. But Pavel Nikolayevich was tormented, no less than by the disease itself, by having to enter the clinic as an ordinary patient, just like anyone else." (Translation by Nicholas Bethell and David Burg).

Five decades later, medicine still faces similar issues. Since few people are ever prepared for a hospitalization, this book aims to provide patients with all the necessary information to make their hospital stay as smooth as possible. It is also intended to serve as a complement to the doctor-patient relationship as it tries to answer questions that a patient or family member might not be able to ask in person. It is my hope this book will dispel, or at least lessen, the feelings of hopelessness and fear that inevitably occur when one is admitted to the hospital. If the information in this book results in a more comfortable, relaxed stay for both the patient and family alike, then it has succeeded.

No matter how complicated the running of a modern health care facility may appear to be, its guiding principles are quite simple. Hospitals function through an uneasy alliance between medicine and business. True, a doctor's primary concern is for the health of each patient, and medicine takes precedence, but there would be no treatment if these facilities were not compensated for their time and services. This symbiosis lies at the heart of American health care. By recognizing the interplay between these functions, a patient will be better equipped to understand the full experience of his or her hospital stay.

From a medical perspective, today's hospitals are designed to treat the sick by hosting a multidisciplinary roster of professionals with varying expertise in the medical arts and healing sciences. Divided into specialized departments that work in close conjunction with one another, these individuals are trained in every aspect of a hospital stay, and serve as the front line against sickness as it manifests in a seemingly infinite number of ways. Still, the wealth of knowledge possessed by this prestigious group is often lost on the typical patient who simply does not understand why he or she happens to be lying in a sterile white bed surrounded by strange people bustling back and forth.

At the beginning of the twentieth century, hospitals were a place where patients went to die. If one had the desire and the necessary resources to survive, treatment took place at home. We should always be mindful just how far medical science has evolved so that we might better understand how the modern hospital operates. In this way, patients are better prepared to appreciate the health care system itself.

On March 23, 2010, President Barack Obama signed the Patient Protection and Affordable Care Act (Health Care Reform) into law. More than ever before in recent history, the topic of health care is in the public spotlight. The legislative changes to our nation's system for delivering health care will continue to unfold in the years to come, and their impact will be better understood over time. The success or failure of health care reform may hinge on how well we as a nation come to understand and appreciate our new system. For better or worse, this education may start unexpectedy during a hospital stay, which often occurs without warning.

The chapters in this book will familiarize you with many of the processes and procedures involved in a typical hospital stay. By explaining each part of your visit in simple terms, from admission to discharge, I have tried to eliminate the confusion and fear that surround any hospital experience. I hope that by doing so, this will free patients to focus their energies on their primary task at hand—getting better.

ABOUT THE AUTHORS

Craig B. Garner, Esq.
Prior to assuming the post of Chief Executive Officer at Coast Plaza Hospital in Norwalk, California, Craig practiced law as an attorney and partner specializing in health care issues. He serves on the advisory board for the College of Osteopathic Medicine of the Pacific, Western University of Health Sciences, and on the Board of Directors for LVS Health Innovations, an evidence-based health management and information technology company focused on helping people create sustainable, active, and healthy lifestyles. Craig is also on the Board of Directors of the Los Angeles Opera and the Board of Visitors of Seaver College at Pepperdine University. He lives in Southern California with his wife Natalya.

Thomas A. Gionis, M.D., J.D., M.P.H., M.H.A.
Dr. Thomas A. Gionis serves as President of Aristotle University, Dean and Professor of Public Health at Aristotle University School of Public Health; he was the Founding Dean and Professor of Law at Aristotle University College of Law. Dr. Gionis is a United States Fulbright Scholar in Law, a current member of the United States Fulbright PEER Review Committee Member/Global-Public Health, and former UCLA Visiting Scholar in Public Health. He also serves as a Visiting Lecturer in the Executive Master of Public Health (MPH) program at the UCLA School of Public Health, where he lectures in Health Law and the Principles of Global Health. Dr. Gionis is a frequent Lecturer in bioethics at international United Nations UNESCO bioethics meetings, and serves as Chairman of the UNESCO International Bioethics Journal Club and Chairman of the Joint Study Committee on Clinical Research Ethics. As Medical Director at Coast Plaza Hospital, Dr. Gionis has been recognized as the youngest individual in the United States to be awarded the medical degree (M.D.); he started medical school at 17, and was awarded the M.D. degree at 21 years of age. He is a published author and the Editor-in-Chief for Royal Society Law Review. He lives and works in Southern California.

Galal S. Gough, M.D., F.A.C.O.G., F.A.C.S.
Currently the Chief of Staff at Coast Plaza Hospital, Dr. Galal S. Gough was a practicing OB/GYN with more than 40 years experience in Southern California. Dr. Gough served as Clinical Professor of Obstetrics and Gynecology at USC Medical Center from 1987-1999, and was an Associate Professor for 17 years before that. He has earned fellowships at the College of Obstetrics and Gynecology, the American College of Surgeons, and the International College of Surgeons. Dr. Gough served on the Medical Board for the State of California between 1984 and 1991 in various capacities, acting both as President and the California representative to the Federation of State Medical Boards. Since 1985, Dr. Gough has served on the California Committee of Bar Examiners, and has held a position as committee member for Moral Character from 2000 to the present. He was appointed Chairman of the Committee of Bar Examiners in 2010. Dr. Gough lives and works in Southern California.

ACKNOWLEDGEMENTS

Hospital Stay is collaboration by a few with the hopes of helping many. The authors would like to thank the following individuals:

COLLABORATION PRINCIPALS

Anne Speedie McDonald: Chief Editor who oversaw all copy and content from cover to cover, start to finish, including the 24 words that form this very acknowledgment.
Tim Barkow: Architect who created the blueprints, converted thought to text, and ultimately built that which is now called Hospital Stay.
Cris Dobbins: Principal Artist who through magical photography and graphic design animated the most visceral of content and brought to life matters sometimes associated with death.

CONTRIBUTORS

Patricia Krueger: Contributor and editorial advisor who, in particular, made complementary and alternative medicine part of our Hospital Stay.
David McCabe: Editorial consigliere who has on occasion ensured my thoughts were presentable to the general public.
Julie Sullivan: Resident academician and professional journalist who contributed throughout the entire collaboration.

Additional gratitude to the physicians, employees, and members of the Coast Plaza Hospital community.

Last, but definitely not least: From me to Natalya: Моя любимая. Мой мир.

— Craig B. Garner

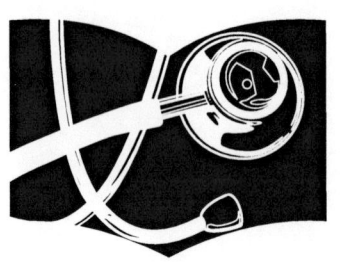 Join the conversation at HospitalStay.com

Want to learn more?
Additional resources and news are available on our website at www.HospitalStay.com. You can also follow us on Twitter @hospitalstay.

Want to buy this book?
You can purchase copies of this book online at: www.HospitalStay.com

Disclaimer: The authors would like to remind you that the information provided in *Hospital Stay* is of a general nature and should never be used to replace the advice of a health care professional. The authors of *Hospital Stay* strongly discourage anyone from "self-diagnosing," including but not limited to any attempt to use or adopt the information in this book as a substitute for a consultation with a doctor. Nothing contained in *Hospital Stay* should be construed as an endeavor to offer or render a medical opinion or otherwise engage in the practice of medicine.

If you believe you have a medical emergency, you should immediately call 9-1-1. If you have any other health concern, or if you have questions regarding your overall health, you should always consult with your doctor or another health care professional. Never disregard or delay medical advice or treatment as a result of something you have read in *Hospital Stay*.

CHAPTER 1: THE PEOPLE IN YOUR HOSPITAL

"No member of a crew is praised for the rugged individuality of his rowing."
– Ralph Waldo Emerson, American author, philosopher, and poet

1

THE PEOPLE IN YOUR HOSPITAL

TOPICS: Anesthesiologist, Case Manager, Doctor of Medicine, Phlebotomist, Radiologist, Registered Nurse

You see them walking the hospital halls or at the reception desk, some in lab coats or surgical masks, others in scrubs. But who are these people, and what do they do? Doctors and nurses play a primary role in treating patients, but so do other health care professionals and administrative staff. Here are some of the people that may impact your hospital stay, whether you meet them or not (in alphabetical order).

Anesthesiologist

An anesthesiologist is a doctor whose primary focus is to administer an anesthetic—a drug that causes unconsciousness or a loss of sensation—to a patient before a medical procedure is performed. After medical school, anesthesiologists complete an accredited residency program in anesthesiology. There are also

Chapter 1: The People In Your Hospital

nurse anesthetists who, after completing basic nursing education and additional anesthetic training, may administer anesthetics, usually under a doctor's supervision.

Case Manager/Care Manager
A medical case manager can be a doctor, a nurse, or a social worker. He or she is the patient's liaison in navigating through health, legal, and/or financial concerns, which almost always play a role in any hospital stay. The case manager works with medical and administrative staff to help identify appropriate providers and facilities throughout treatment and ensures that resources are used in a timely and cost-effective manner. In other words, case managers try to get you more bang for your buck and see that your medical providers are fairly compensated.

Certified Nursing Assistant (CNA)
Under the supervision of a registered nurse (RN), certified nursing assistants provide bedside care and perform routine nursing procedures—like starting an IV or taking blood pressure—enabling the RN to focus on other responsibilities. Although CNAs do not administer certain medications or assist in preparation of surgery rooms, they still watch over patients and handle many of the RN's daily tasks. Good communication between CNAs and RNs is vital for consistent and competent patient care.

Certified Registered Nurse Anesthetist (CRNA)
As the name suggests, a certified registered nurse anesthetist is a licensed registered nurse with additional graduate-level education in the field of anesthesiology. Depending upon state law, CRNAs may work independently or under a doctor's supervision. To assist an anesthesiologist during operations, or if a patient is at risk for complications, surgical teams may include a CRNA to help administer and monitor sedation and pain control.

Chaplain

Usually a member of the clergy, a chaplain is typically a priest, pastor, rabbi, or ordained deacon. In the United States, most health care chaplains are educated through the Association for Clinical Pastoral Education and may be certified by a national association of chaplains. Often hospitals rotate chaplains of varying faiths, and patients can request visits from any chaplain on duty or clergy representing their own faith.

Clinical Laboratory Scientist (CLS)

Found in every hospital laboratory, a clinical laboratory scientist performs a wide range of lab tests, from simple blood analyses to complex DNA tests. They provide doctors with information needed to diagnose and treat any number of ailments, including heart disease, cancer, diabetes, and certain types of infection. These health care professionals are often well-versed in the latest biomedical technology, using computers, microscopes, and specialized electronic equipment to perform tests on bodily fluids and tissue specimens. They often work in conjunction with doctors and nurses to assess a patient's physical health.

Dietitian/Registered Dietitian (RD)
Dietetic Technician, Registered (DTR)

Clinical or therapeutic dietitians ensure that hospital patients receive nutritious food according to their medical needs. By developing individual plans, dietitians provide "medical nutrition therapy," including tube feedings (called enteral nutrition), intravenous feedings (called parenteral nutrition), diets, and education for patients and family members. Some are also responsible for food planning and service for the whole hospital. To use the label "Registered Dietitian," they must meet strict educational and professional prerequisites, regulated by the American Dietetic Association (ADA), and pass a national registration examination.

CHAPTER 1: THE PEOPLE IN YOUR HOSPITAL

Some RDs or DTRs call themselves nutritionists, but this title is not regulated.

Doctor of Medicine (MD)

Also known as a physician, a doctor of medicine is a licensed health care practitioner who studies, diagnoses, and treats disease in the human body. Having earned a medical degree from an accredited medical school, MDs may practice medicine, prescribe drugs, and order tests for patients. In a hospital setting, doctors work closely with nurses, technicians, and other health care professionals to assess a patient's health and diagnose any disease-related issues. Ultimately, MDs bear the responsibility for the safety and welfare of their patients.

Doctor of Osteopathic Medicine (DO)

Though they undergo similar training as MDs, Doctors of Osteopathic Medicine incorporate a more holistic approach to health care. Regarding the human body as an integrated whole, DOs focus on the interconnection of nerves, muscles, tissues, and organs on the overall health of the patient. Through the use of osteopathic treatments such as the application of pressure to specific points within the musculoskeletal system, DOs seek to realign the body's integral organs and tissues into a healthier position so the body is better able to heal itself naturally. In a hospital, the primary treating physician could be an MD or a DO. Both are licensed to diagnose and treat diseases.

Insurance Verifier

Confirming whether patients have insurance and what their plans cover is no small task. Insurance verifiers must simultaneously juggle hundreds of patients and dozens of insurance companies, all of which include varying plans, benefits, and co-pays/deductibles. Each insurance plan has different rules governing which procedures it will cover, how much it will pay, and when authorization is required. Managing all of these aspects, as well as the paperwork involved with getting patients the coverage they need, keeps the insurance verifier busy.

Interpreter
According to a survey by the Health Care Interpreter Network, more than 70 percent of health care providers say that language barriers compromise patients' understanding of their disease and treatment advice, increase the risk of complications, and make it harder for patients to explain their symptoms. For this reason, many hospitals have interpreters available to translate such languages as Spanish, Cantonese, Mandarin, Vietnamese, Lao, Mien, Thai, Cambodian, Hmong, Korean, Russian, Farsi, Armenian, Tongan, and Hindi.

Licensed Practical Nurse (LPN)/Licensed Vocational Nurse (LVN)
Whether in clinics, hospitals, or private health care, licensed practical nurses often work under the supervision of doctors and registered nurses. They are often responsible for taking and monitoring vital signs, gathering a patient's health history, preparing and delivering injections, collecting lab samples, and teaching patients about positive health habits. In many settings, they are responsible for supervising nursing assistants and aides.

Nurse Practitioner (NP)
Though the rules governing the nurse practitioner vary by state, NPs are registered nurses who hold master's or doctoral degrees in advanced nursing education, combined with training in the diagnosis and management of many medical conditions. Depending upon location, NPs may work independently of doctors, as they are trained to treat a variety of both acute and chronic conditions. Their role in a hospital includes the taking of personal health

Chapter 1: The People In Your Hospital

histories, giving physical exams, ordering tests and/or therapies, and prescribing certain drugs.

Phlebotomist
Phlebotomists are health care professionals trained in drawing blood for the purpose of analysis. While some hospitals allow on the job training, most phlebotomists complete four months in a career center or trade school or one year in a community college. This training includes a hospital clinical rotation and cardiopulmonary resuscitation (CPR), as well as anatomy, patient interaction, Universal and Standard Precautions, blood collection techniques, and legal aspects.

A phlebotomist must possess the ability to take samples from diverse patient groups such as the elderly and infants while minimizing the risk of transmission of disease through contact with bodily fluid. Phlebotomists store and label drawn blood, while placing those patients who fear needles at ease.

Physician Assistant (PA)
A physician assistant is a licensed health professional who provides a broad range of health care services under the specific supervision of a licensed physician. These include the giving of physical exams, ordering tests, diagnosing illnesses, and assisting in certain surgeries. A PA's role often varies with experience, training, and state law. Though the supervising physician is ultimately responsible for the health and safety of the patient, PAs make many important medical decisions and are able to prescribe medication in all 50 states.

Psychiatrist
While psychologists focus primarily on a patient's emotional state, psychiatrists are medical doctors (MDs) who hold specialized degrees in psychiatry and the treatment of mental disor-

ders. Their areas of expertise allow them to address the mental and physical state of a patient by combining psychological and biological means of evaluation and treatment. Unlike most psychologists, a psychiatrist may issue prescription drugs, order laboratory tests and neuro-imaging scans, and conduct thorough physical examinations.

Psychologist

Psychologists are mental health professionals who hold a license to practice psychology and provide psychotherapy. Though there are many types of psychologists, the most common in a hospital setting are clinical and counseling professionals. It is their job to administer and interpret psychological tests to ascertain a patient's well-being and current state of mental health. In many states, psychologists cannot legally prescribe drugs, so they often work with a doctor to determine the best course of action for the patient. A psychologist may also offer counseling or psychotherapy, depending upon the patient's situation.

Radiologic Technician and Technologist

These health care professionals oversee most radiological diagnostic tests. Technicians coordinate basic imaging tests, such as X-rays. In preparing any patient, technicians explain the procedure and position the patient and equipment in order to get the best possible image. They also try to minimize the patient's exposure to radiation by using protective devices and taking images as quickly as possible.

Technologists, on the other hand, assist with more advanced tests, such as CT scans, MRIs, and mammograms. Technologists rely closely on a doctor's order for the imaging study. Given the higher levels of radiation that some of these tests may emit, technologists always work to prevent unnecessary exposure to patients, themselves, or anyone else in the hospital. Sometimes,

Chapter 1: The People In Your Hospital

a technologist will also introduce a medical contrast, usually by injection, into the patient. Contrasts are used to highlight certain parts of the body in order to obtain a better image.

Radiologist
Radiologists are licensed doctors who use medical imaging technologies to diagnose or treat disease. They originally dealt solely with X-ray machines and other radiological devices (hence their name), but today they also work with advanced imaging technologies such as ultrasound, computed tomography (CT), and magnetic resonance imaging (MRI). Usually, a technician performs the imaging procedure, and a radiologist reads the results and delivers them to your attending doctor. Radiologists can also assist minimally invasive surgeries by guiding imaging technology. After completing medical school, radiologists must complete five years of training, including a diagnostic radiology residency.

Registered Nurse (RN)
Registered nurses are the largest group of health care workers in the United States. In a hospital, their job is to provide direct treatment and care for patients, including monitoring vital signs and preventing a variety of illness-related deaths such as infection, cardiopulmonary arrest, and other serious complications. RNs are often seen as intermediaries between doctors, technicians, and the patient, though they may also act as supervisors and managers for their department, enforcing policies, responding to emergencies, overseeing CNAs, and determining budgets. In the U.S., all RNs must graduate from an accredited nursing program, though many pursue advanced degrees in both medicine and management.

Social Worker
In a hospital setting, clinical social workers work with case managers to provide services such as counseling, referrals to commu-

nity agencies, help with legal and financial issues, and support for patients and families in dealing with the stress associated with managing illness. They typically have a master's degree in social work and aim to assess the patient's and family's psychosocial functioning, intervening as necessary. They tend to work with professionals of other disciplines, such as medicine, nursing, and occupational and recreational therapy. Their approach involves considering the whole individual (including biological, psychological, sociological, familial, cultural, and spiritual aspects) within the context of the current situation.

Surgeon
In medicine, a surgeon is a licensed physician who performs surgery. Surgeons treat a broad category of diseases, injuries, and deformities, some by operating (e.g., open heart) and others by physical manipulation (e.g., an orthopedic surgeon "reducing," or returning a dislocated shoulder to its normal position). Despite its name, general surgery is a surgical specialty that focuses on surgical treatment of abdominal organs, including the esophagus, stomach, colon, liver, gallbladder, and bile ducts. Depending on the availability of head and neck surgery specialists, it may also include the thyroid gland and hernias. Following high school, it takes approximately 13 years to become a fully licensed general surgeon (four years of undergraduate training, four years of medical school, and five years of residency).

CHAPTER 2: THE EMERGENCY ROOM

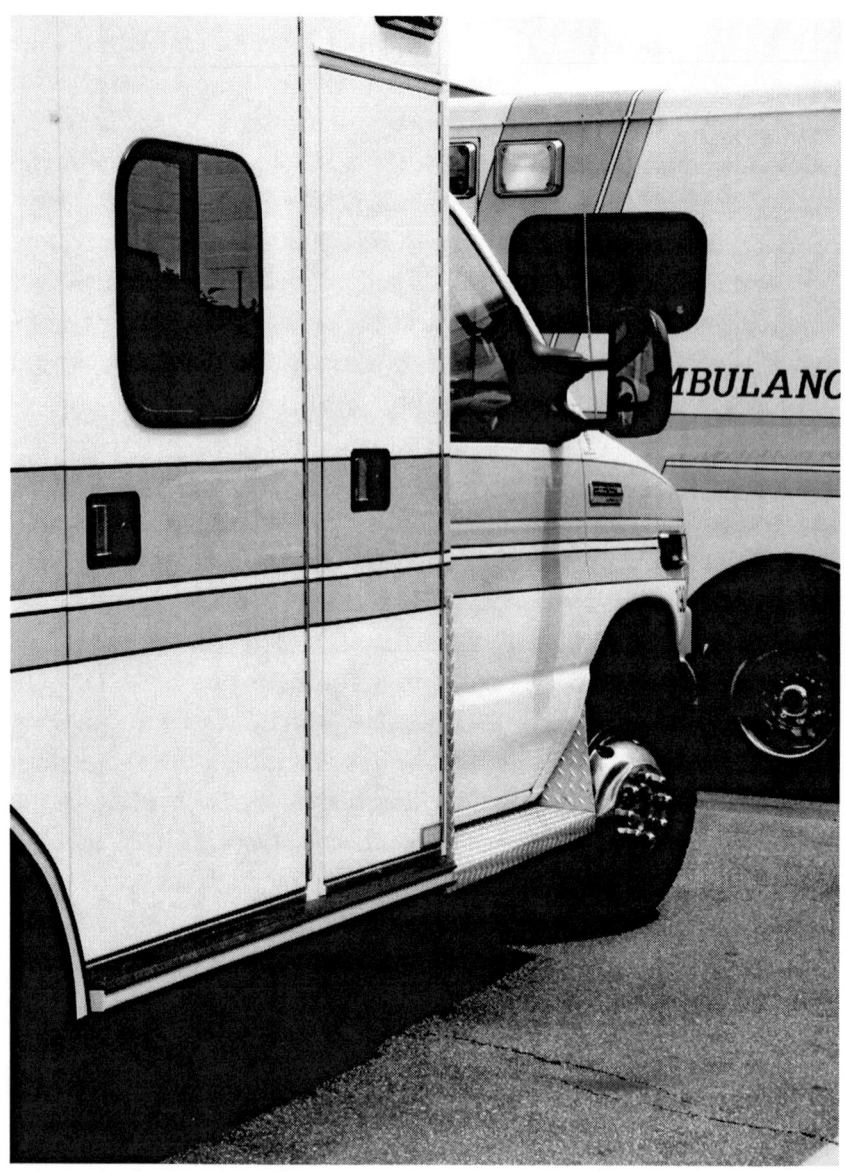

"The first thing about being a patient is you have to be patient."

— Dr. Oliver Saks, British neurologist, author, and patient

HOSPITAL STAY

2

THE EMERGENCY ROOM

TOPICS: Hospital Codes, Insurance, Priority for Admission, Trauma Center, Triage, Urgent Care, Waiting

You probably have an idea as to what happens in a hospital emergency room (ER) from what you've seen on TV. An unconscious patient is wheeled in by paramedics, at which point doctors with spectacular diagnostic skills immediately treat the patient, almost always successfully. ERs in the real world tend to work much differently. While you may see a doctor who treats your emergency, you will likely first have to wait for what seems like a long time as you fill out scores of paperwork.

Check In, Sit Down, Wait
Unless you are unconscious or experiencing an immediate emergency, you will first be asked to sign in at the ER front desk and state the nature of your illness. You will be asked whether you have health insurance, and if the answer is yes, you will be requested to

Chapter 2: The Emergency Room

show your health insurance card for billing purposes. Next, you will likely be told to take a seat and wait for your name to be called.

As you wait in the ER, you may spend hours next to people you wouldn't ordinarily encounter in your everyday life. Like jury duty and traffic school, the ER breaks down social and economic barriers, placing the shiny-shoed businessman next to the crackhead with a knife wound and the mother with infant next to the mentally ill homeless person in search of medication and a free meal. Despite the differences you'll see in the ER waiting room, patients are given priority according to the nature of their condition, not social or economic status.

Who Is Seen First?

Determining patient priority is not always easy. In the ER, everyone believes his or her case is the worst, and that he or she has been waiting the longest to be treated. Due to a high volume of patients and limited staff, ERs use a process called "triage" (pronounced "TREE-ahj") to assess the nature of each injury or illness. Triage, from the French "to sort or separate," is usually done upon a patient's arrival by a trained hospital health care professional. This person will evaluate your condition, noting any changes that may occur during the waiting period, and compare this information with those of other patients seeking treatment concurrently. This usually determines priority for admission. ERs may use the following five-scale scheme to rate major to minor injuries:

> **Expectant**
> These patients present with such severity of injury, such as deep trauma, large-scale burns, or cardiac arrest, that they are not expected to survive. They are taken to a holding area where their suffering is alleviated with painkillers.

Immediate
These patients are likely to survive if given immediate lifesaving intervention, often in the form of surgery. They are the highest priority in any ER, as any delay in treatment will result in swift decline.

Observation
These patients are currently stable but in need of regular observation by trained staff to monitor their progress. They will require further medical care outside the ER, and are normally seen immediately if possible.

Wait
These patients are in need of medical attention within the next few hours or days, but their conditions are not immediately life threatening. They may be forced to wait indefinitely, as priority will be given to any patient, including those who arrive after, presenting with more severe injuries.

Dismiss/Discharge
These patients have only minor injuries that do not require a doctor's care. Home first aid and maintenance are sufficient.

Why Is the Wait So Long?
The time you spend waiting in an ER may depend on numerous factors, not just the triage process. Federal laws mandate that ERs treat all patients who walk through their doors—24 hours a day, seven days a week, without exception and regardless of ability to pay. Although many patients often feel frustrated and without equal access to medical treatment in the event of sudden illness or catastrophe, few people realize that the ER is open to everyone. Like it or not, this is another reason you may find your local ER packed with people.

Chapter 2: The Emergency Room

How Did It Come to This?
As its name suggests, the emergency room was initially created to treat medical situations such as life-threatening illnesses and unforeseen traumas. As health care in the U.S. has evolved, however, so too has the role of the ER. Over time, as the nation's uninsured population has increased, the ER has largely borne the brunt of the overflow, and now serves as a catch-all for the underprivileged and uninsured sick, while retaining its original duties in providing emergency medical assistance.

How Do I Shorten My Wait Time?
The easiest way to spend as little time in your local ER waiting room is to only visit the ER in the event of a life-threatening emergency where you need immediate medical assistance. In short, don't go if you don't have to. Illnesses such as the flu and injuries such as a sprained ankle or a cut that may need stitches can be treated promptly in your primary care physician's office during business hours. If you become sick after hours, your doctor's emergency line may be able to give advice over the phone or refer you to an urgent care center that treats emergencies that are not life-threatening. (You might also consider visiting an urgent care center if you are injured or fall sick while traveling.)

Getting seen at an ER is dependent not only on the severity of your condition, but on the conditions of other waiting patients as well. Asking yourself if you really need ER treatment is better for everyone. The more focused an ER is, the more successfully it can do the job it was created to do, and the better prepared it will be to serve the community in the event of an actual emergency. After all, that's what the "E" in ER stands for.

DECIPHERING HOSPITAL CODES

Hospital codes are short phrases—usually named after colors—that health care workers use to talk about serious issues quickly and clearly. Codes are an easy way to ensure all workers are using the same terms—an especially important task because hospitals employ large staffs that work in varying shifts. By using a common language, workers can respond to emergency situations quickly, reduce errors, and improve care at the same time. Codes are also useful in alerting staff without alarming visitors and patients.

While there is no national standard for hospital codes in the United States, these are some of the most commonly used hospital codes and their meanings:

Code Amber is named for abducted child Amber Hagerman, whose murder inspired the nationwide AMBER Alert system (America's Missing: Broadcast Emergency Response). It is most often applied to suspected child or infant abduction. In some hospitals, however, Code Amber indicates a theft or armed robbery in progress within the facility.

Code Black is used by U.S. military hospitals and certain civilian hospitals to denote mass casualties caused by an epidemic or threat to public health. In the Midwest, Code Black often indicates impending severe weather, and it may also serve to alert employees of a bomb threat, child abduction, or nuclear attack. In Alaska and Georgia, Code Black conveys that a patient has been pronounced dead.

Code Blue is used to indicate when a patient needs immediate life-saving assistance, usually as the result of cardiac arrest. All available employees are required to respond immediately. In an effort to provide more specific information, the Hospital Association of Southern California recommends Code Blue be used only

Chapter 2: The Emergency Room

for adult cases, with Code White referring to medical emergencies involving children.

Code Brown may refer to a severe weather warning, a medical gas emergency, a missing adult within the hospital, or a patient with uncontrolled bowel issues (a health hazard). Tampa General Hospital uses Code Brown to call for autopsy.

Code Gray, according to the Hospital Association of Southern California, is used to convey that a violent yet unarmed person is loose within the hospital. It may also refer to a patient having a violent psychotic episode who is in need of restraint and/or medication. It is sometimes used to denote severe weather or emergency response for inpatient stroke.

Code Green has a number of meanings across the U.S. Although it is most often used to indicate a situation where evacuation is in order, such as an internal or external disaster, it may also refer to an armed and combative person within the hospital. Code Green may also be used to call a fire drill, convey a difficult delivery in Obstetrics, or to sound the All Clear after an emergency, indicating that personnel should resume their normal duties.

Code Orange may indicate a bomb threat, a prisoner escape, or warning of an incident involving hazardous material within the hospital. It is also used to alert emergency medical teams of a patient whose health is rapidly declining.

Code Pink usually refers to biohazardous contamination of a patient or employee. It may also indicate imminent birth with no physician present, an infant or child abduction, or a nurse being harassed or abused by a physician.

Code Red most often indicates a fire in the building, though it may be used to alert staff to the impending arrival of a patient with burn injuries. It may also refer to an incoming life-threatening trauma, or that an external disaster has occurred and many casualties are on the way.

Code Silver refers to a combative person with a lethal weapon. In most Southern California hospitals, it often indicates a violent situation and potential hospital lock down.

Code Yellow, according to the Hospital Association of Southern California, is used to indicate a bomb threat within the hospital. In many trauma centers, it is used to alert the emergency response team that a patient is in imminent danger of death. It may also indicate a severe weather alert, depending on location.

Code White most often denotes a pediatric medical emergency, though it may also refer to a power or utility outage, a natural disaster requiring mass evacuation, or impending severe weather.

CHAPTER 3: INTRAVENOUS THERAPY

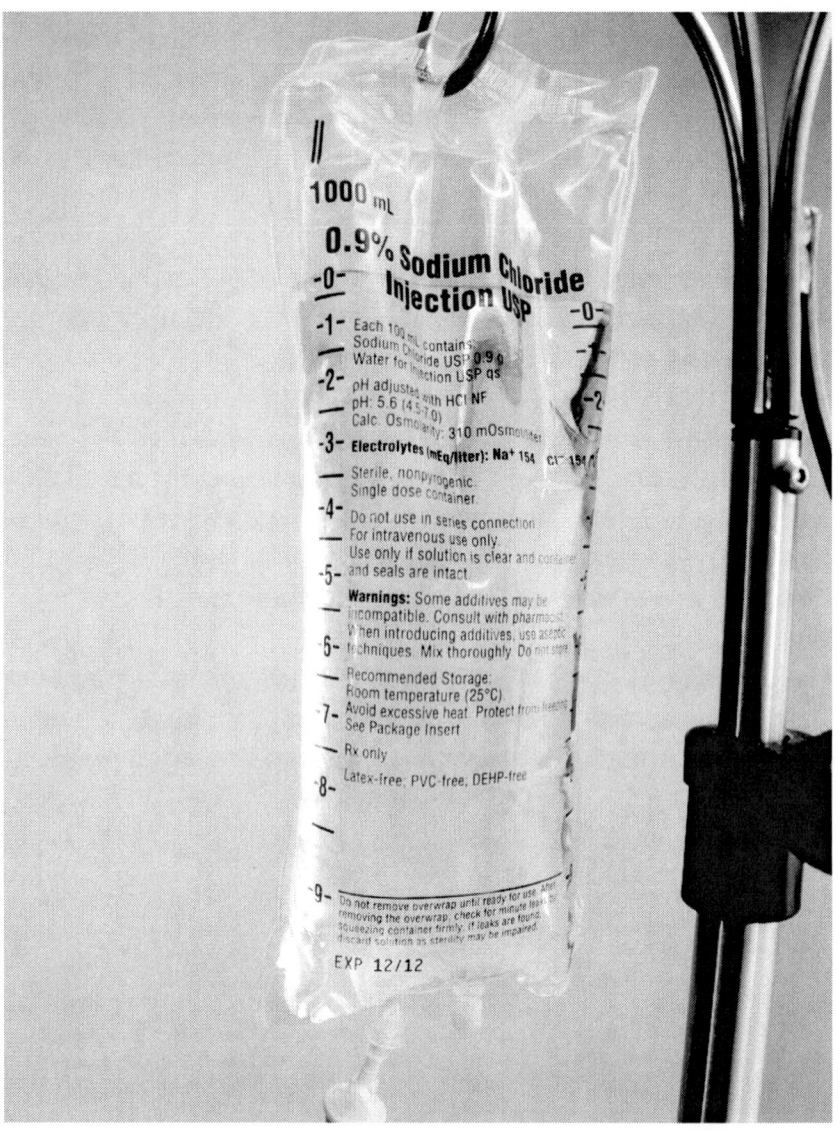

"Medicine when injected directly into the veins often works more swiftly and successfully than medicine given through the stomach."
— Dr. W. Forest Dutton, Medical Director for the Hospital of the University of Pennsylvania, July 14, 1924

3

INTRAVENOUS THERAPY

TOPICS: Artery, Catheter, Central IV, Extravasation, Heparin, Implantable Port, Infiltration, Peripheral IV, Phlebitis, Vein

If you don't like needles, you're not alone. According to a 2001 Gallup poll, one in five Americans say needles scare them. So why do most hospital visits require them? Intravenous access (meaning through the veins) is the simplest, most efficient way to deliver necessary fluids and medications throughout the body. In most cases, IV therapy works faster than inhaled or swallowed medications, and it can better deliver measured doses over a period of time.

Your doctor might recommend IV therapy to replace fluids and electrolytes for various conditions. IV therapy can help with dehydration by sustaining fluid and electrolyte balances when you cannot maintain appropriate levels, and it is useful for giving a patient Total Parenteral Nutrition, which provides necessary calories for energy. It is also used to replace blood productions in the

CHAPTER 3: INTRAVENOUS THERAPY

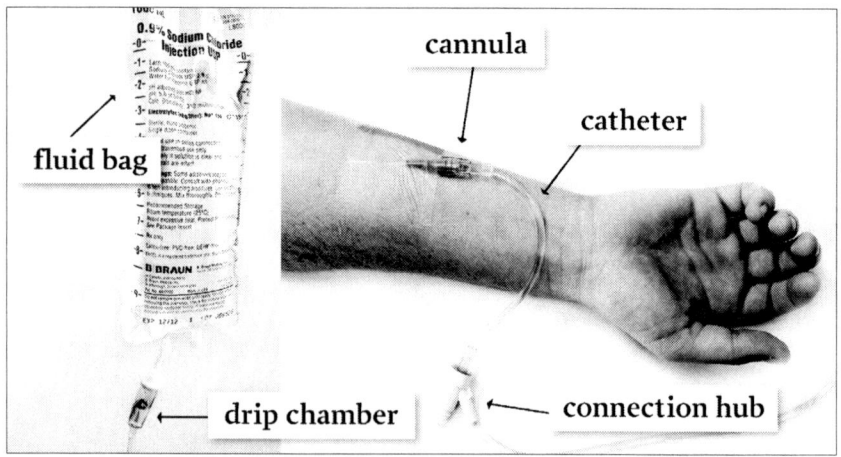

A typical IV is made of just a few simple pieces.

body, help keep veins accessible for blood draws, and to administer medication.

How an IV Works

Knowing why you need IV therapy makes it a lot easier to roll up your sleeve, so let's explain how it works. What doctors and nurses call an IV is really just a long rubber or silicon tube, called a catheter, with a small needle attached at one end and a fluid-filled bag at the other. External tubing that joins through a connection hub allows one or more points of access to the site. (See photo above.)

To start IV therapy, a nurse or other hospital worker takes your arm and inspects it, looking for a "good" vein, meaning one that bulges or puffs out for optimal intravenous access. If a good vein is difficult to find, the correct use of a light tourniquet (such as an elastic band) may be appropriate. A warm, damp cloth may be used to increase circulation and expand the target vein.

After a good vein is located, the skin is cleaned with a special solution (sometimes dark brown in color or as clear as alcohol) to prevent bacteria or foreign materials from coming into contact with the skin that covers the target vein. Then the IV catheter of the needle is inserted through the skin, into a vein. This is generally quick and painless, but it may feel uncomfortable at first.

Sometimes obtaining access can be difficult due to the constriction of veins and lower fluid volume. This can result from factors such as age, severe dehydration, fatty tissue underneath the skin or substantial lack thereof, or severe cold conditions.

Once the skin has been pierced, the catheter is inserted into the vein and the needle retracts. Then the contents of the bag flow through the catheter and the needle into the bloodstream. That's it. Because our blood circulates throughout our entire bodies, IV therapy is usually the fastest way to get medicine where we need it most. And because the catheter allows for continuous access, nurses can control the rate and amount of medicine as needed.

IV Access

There are two main types of venous access: peripheral IV lines and central IV lines. When your doctor chooses which type of IV to use, consideration is given to many factors, such as your personal comfort, age, the intended use of the IV, whether you are right- or left-handed, and even certain medical conditions and/or complications. Your physical stature may also play a role, dictating the size of the catheter, the status of blood circulation throughout the body, and whether access to certain veins is possible.

Hospitals try to use the smallest possible tube (called a "cannula") to minimize trauma to the vessel and allow the flow of blood around the area to remain uninterrupted. In an emergency, however, a larger catheter may be necessary for rapid infusion into the system.

If you need IV therapy during your hospital stay, you'll most likely receive the least invasive type: a peripheral IV line. These lines are the most common, generally changed every 72 to 96 hours to prevent risk of insertion-site infection. They are administered through any vein not found in the abdomen or chest, usually in the

CHAPTER 3: INTRAVENOUS THERAPY

> **WHAT IS HEPARIN?**
> Heparin is a naturally occurring substance known as a blood thinner. While it does not actually thin the blood itself, heparin minimizes coagulation to prevent blood clots from forming within the blood stream. It also forms an anti-coagulant covering on various medical devices and catheters that are used internally within the human body.
>
> Heparin is one of the oldest drugs in popular use, and is made by pharmaceutical companies who manufacture it from the surface linings of slaughtered animals (most commonly from the intestines of pigs or the lungs of cows).

forearm or hand, but also in the foot or lower leg. These peripheral sites are furthest from the heart, and often provide easy access. Should you develop an infection as a result of an IV, it is best that the infection starts as far away from the heart as possible.

Central lines are more invasive and less common than peripheral lines. They are used only in serious medical conditions or when large volumes of blood, fluids, or serious medications must be given. Due to increased risk for infection, bleeding, and gas embolism (gas bubbles in the bloodstream that, if large enough, can stop the flow of blood in the body), central venous access is reserved for special circumstances such as long-term antibiotic therapy, dialysis, chemotherapy, or for administering certain medications that can cause irritation of smaller peripheral veins.

Most central lines are placed using local anesthesia to limit discomfort. This procedure may be conducted in an operating room. If necessary, an X-ray may accompany this procedure to confirm proper placement of the catheter tip.

Central IV lines include peripheral inserted central catheters (PICC), central venous lines, and implantable ports. Central lines are used for direct access to larger veins in a patient, such as the superior vena cava (the vein near the neck and chest area), the inferior vena cava (the vein near the abdominal area extending toward the heart), the jugular (the primary vein in the neck), or the femoral vein (the vein near the groin area, just before the inferior vena cava). These insertions typically end just before entry to the right atrium of the heart.

Central Lines
The difference between types of central lines is based primarily on the catheter's route as it makes its way to the desired vein.

PICC Lines
A peripheral inserted central catheter (PICC) line is inserted at the elbow and "threaded" up through the vein to a position in the superior vena cava, just above the heart. While this procedure sounds uncomfortable, it is actually no more painful than inserting a peripheral IV line, and it can be administered by a doctor, nurse, or physician assistant.

Once in place, the line is often verified by an X-ray. These lines are convenient for patients requiring long-term intravenous access who are permitted to leave the hospital. Proper care, such as keeping the dressing clean and dry, must be maintained to prevent infection at the site.

Fortunately, because a PICC line is predominantly internal and may remain in place for two to six months, patients should be able to continue normal activities simply by flushing the point of entry with saline (a salt and water solution) or heparin (an anticoagulant or "blood-thinning" agent) once a day or after each use.

Permanent Central Catheters
These tunneled catheters are central lines that enter the body on the right side under the collarbone and above the nipple. They may be placed for permanent use, so patients who require IV medications on a daily or frequent basis are typical candidates

Chapter 3: Intravenous Therapy

for this procedure. Central catheters allow the patient to resume normal activities, and once taped to the chest they are barely noticeable. With a tunneled catheter, you can administer medication yourself at home. Maintenance of the catheter requires you to flush the line with either saline or a mix of saline and heparin every one to seven days, depending on your doctor's instructions.

These lines are called tunneled catheters because they tunnel through approximately 6 inches of tissue to the superior vena cava. At the site of entry, the catheter has a protective cuff that is surrounded by skin in order to guard the body against dangerous infections such as a bacterial invasion into the bloodstream. Central catheters are inserted surgically while the patient is under appropriate anesthesia. The procedure usually causes soreness for a few days, but the area should heal within six weeks.

Implanted Catheter Ports
Like a permanent central line, an implanted catheter port allows mobility so you can go home, but because its connector, or "port," is implanted under the skin, a health care professional should perform treatments. Made of titanium or polysulphone (a tough, heat-resistant plastic), these lines are surgically implanted under appropriate anesthesia.

Initially, a port is inserted with a rubber seal covering that is implanted under the skin of a patient's chest, leading to the vein beneath the collarbone. Once in place, a small needle is then used to pierce the rubber seal and access the port. The cover is specifically designed to reseal itself once the needle has been removed, and can accept hundreds of needles over the course of its lifetime.

These ports carry a lower risk of infection than PICCs, and can remain implanted for long periods of time if flushed monthly with an anti-coagulant. This especially benefits patients requiring medication administration more than once a week on a regular basis. Over a period of time the entry point may become callus due to the thickening of the external layer of skin, resulting in little if any pain at the point of entry. Advantages to the implanted port include less maintenance at the site of implantation, added protection from the skin barrier when not in use, and minimal visibility and obtrusiveness.

Understanding the Risks

The utility of intravenous therapy has remained constant since Dr. W. Forest Dutton's discovery in 1924 that IVs are more effective at delivering medication than oral therapy (pills, liquids, etc.). As with all medical procedures, however, there are some things to keep in mind. Your doctor will weigh the risks versus benefits when deciding which type of IV catheter suits your needs. Generally, he or she will consider the likelihood of infection, inflammation, and infiltration.

Any break in the skin carries a risk of infection, even a skinned knee. Although IV insertion is usually a sterile procedure, skin-dwelling organisms may enter through the insertion site around the catheter. Bacteria may also be accidentally introduced inside the catheter from contaminated equipment. Moisture introduced to unprotected IV sites through washing or bathing substantially increases the risk of infection.

Infection of an IV site is usually local, causing easily visible swelling, redness, and fever. If unwanted bacteria do not remain in one area but spread through the bloodstream, the infection is called septicemia (colloquially called blood-poisoning), which can be rapid in onset and potentially life-threatening. An infected central IV poses a higher risk of septicemia, as it delivers bacteria directly into central circulation.

Phlebitis, or inflammation of a vein, is another risk. Thrombophlebitis occurs when a vein's inflammation is caused by a blood clot. Leg veins are most prone to thrombophlebitis, though it may also occur in an arm. The clot in the vein causes pain and irritation

Chapter 3: Intravenous Therapy

and may block blood flow throughout the body. Phlebitis can occur in either the superficial or deep veins.

Superficial phlebitis affects veins on the surface of the skin. This condition is rarely serious and usually resolves rapidly with proper care. Deep vein thrombophlebitis affects the larger blood vessels deep within the legs, and is of greater concern. Blood clots (thrombi) can form, which may break off and travel to the lungs, causing a potentially life-threatening condition called pulmonary embolism. Since patients with superficial phlebitis may also get deep vein thrombophlebitis, a medical evaluation is essential if the patient is at risk.

Infiltration, another risk, occurs when IV fluid accidentally enters the surrounding tissue rather than the intended vein. Though not often painful, infiltration is characterized by a coolness and pallor to the skin, as well as local edema. Treatment involves removing the intravenous access device and elevating the affected limb so that the collected fluids can drain away. Infiltration is one of the most common adverse effects of IV therapy and is not often serious unless the infiltrated fluid is a medication that proves damaging to the surrounding tissue, in which case the incident is known as extravasation.

Extravasation is the accidental administration of IV-infused medicinal drugs (or the IV fluid itself) into the surrounding tissue, either by leakage (e.g., due to brittle veins in very elderly patients), or direct exposure (e.g., when a needle punctures the vein so that the infusion goes directly into nearby tissue). Extravasation of medicinal drugs during intravenous therapy is a preventable side-effect. In mild cases, extravasation may cause pain, reddening, or irritation of the limb with the infusion needle, while severe damage may include tissue necrosis. In extremely rare cases, it may lead to loss of the infected limb.

There are also risks surrounding the administration of a solution that is either too-diluted or too-concentrated. This can disrupt the patient's balance of sodium, potassium, magnesium, and other electrolytes. This is why hospital patients normally receive blood tests to monitor these levels.

There is nothing new about IV therapy. With benefits and risks long-studied and well-known, IV therapy is a procedure in which nearly every health care worker receives training. While it may cause some discomfort in patients, there is no better way to introduce fluids and medication into the human body.

CHAPTER 4: VITAL SIGNS

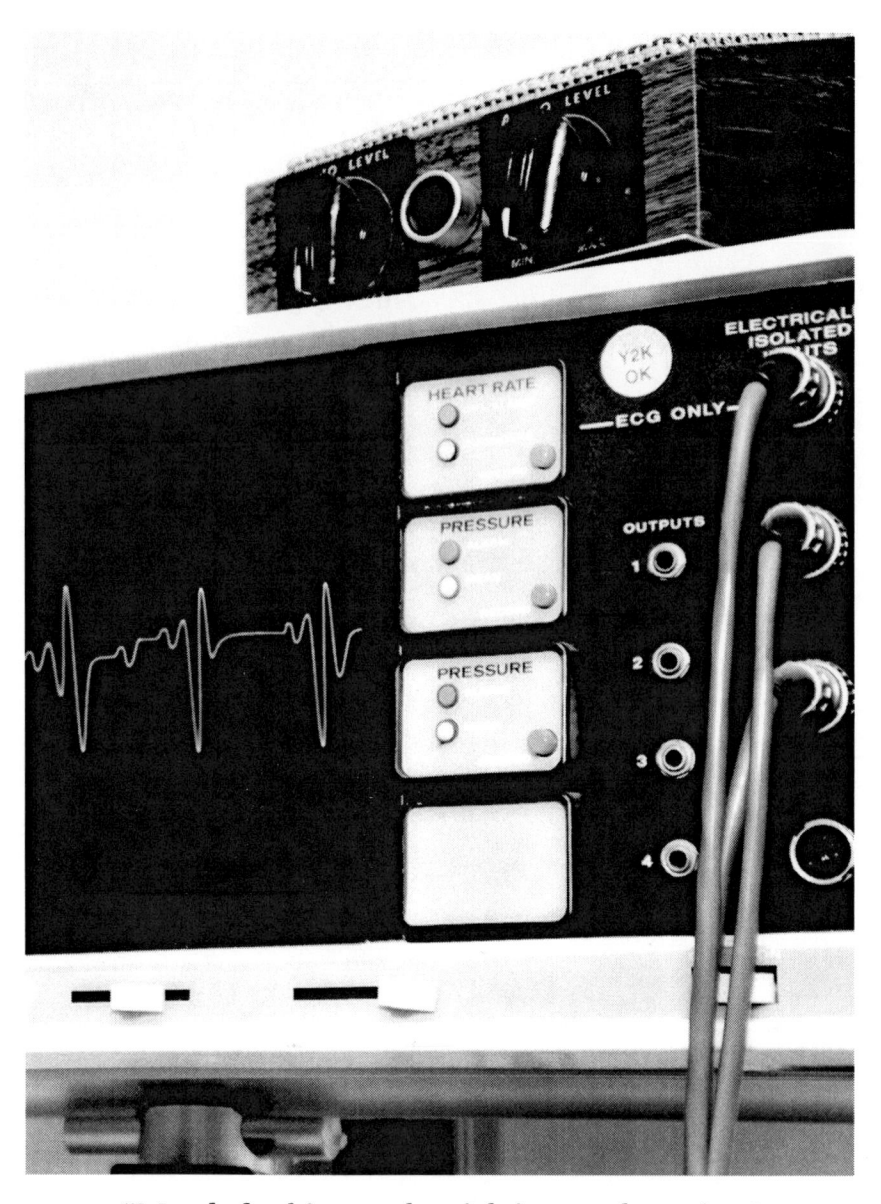

"Merely looking at the sick is not observing."
— Florence Nightingale, British nurse, author,
and patient advocate

HOSPITAL STAY

4

VITAL SIGNS

TOPICS: Blood Pressure, Body Temperature, Hypertension, Pulse Rate, Respiratory Rate, Systolic and Diastolic Pressure

Almost every time you visit the doctor someone will take your blood pressure, pulse, temperature, and/or respiratory rate. These measurements, called "vital signs," are a quick, efficient gauge of overall health. They're called "vital" for a reason—they are vital in helping the doctor evaluate your stability and safety. Used to measure your initial condition and monitor changes throughout treatment, these signs play an important role in just about any medical setting, from hospital emergency services to a simple check-up at your doctor's office.

Normal ranges for the four main vital signs vary in relation to your sex, age, weight, level of activity, and physical condition. For example, a thin track star is likely to have lower blood pressure than a slightly overweight 50-year-old office worker. Included on the following page are basic descriptions and average ranges.

CHAPTER 4: VITAL SIGNS

Vital Sign Ranges

	Low	Normal	High
Blood Pressure	100/60 mmHG or lower	120/80 mmHG	130/85 mmHG or higher
Pulse Rate[1]	60 bpm or lower	60 – 100 bpm	100 bpm or higher
Respiratory Rate[1]	12 bpm or lower	12 – 18 bpm	25 bpm or higher
Temperature[2]	96.8° F (36° C) or lower	98.6° F (37° C) ± 1° F	101° F (38.3° C) or higher

---

[1] Based on a healthy adult over 18. These rates decrease as we age, so infants, children, and teenagers have higher rates.
[2] Based on an oral thermometer reading. Rectal, ear, and armpit readings may be half a degree or so higher.
This information is not intended to replace the advice of your medical provider.

Blood Pressure

Blood pressure refers to the force or "pressure" at which blood is moving throughout the body. It is an indicator of cardiovascular health and an important tool in diagnosing high blood pressure. Also known as hypertension, chronic high blood pressure is directly related to increased risk of heart attack, stroke, and death.

Blood pressure is typically measured using a cuff, which tightens around your upper arm, and a stethoscope, which your doctor or nurse presses to your arm to listen. It can also be taken using an electronic monitor, or invasively through an arterial line. These tools measure the force at which your blood is circulating.

Every time the human heart beats, pressure is exerted upon the walls of the arteries. The measurement of this force is recorded in two separate numbers. The first, systolic pressure, monitors the pressure within the arteries as the heart contracts and pumps blood throughout the body. The second, diastolic pressure, is measured between beats, as the heart relaxes and fills with blood, reducing pressure on the artery walls. These two figures, systolic and diastolic, are reported in millimeters of mercury (mmHG) because the tools once used to measure blood pressure contained mercury.

(Today, nearly every tool containing mercury is being replaced by a mercury-free substitute in many U.S. hospitals.)

If your blood pressure reading is "one-twenty over eighty (120/80)," you are within the normal, healthy range. Our blood pressure fluctuates throughout the day, reacting to variables that include fatigue, stress, emotion, and physical activity. Athletes and children typically have lower blood pressure, while healthy adults can post higher readings in the morning or after eating salty, fatty foods. It is also normal for average blood pressure readings to increase slightly as you age.

Pulse

Your pulse rate, also sometimes called "heart rate," refers to the number of times your heart beats per minute (bpm). It is most commonly measured on the neck or wrist, but can also be taken behind the knee, on the inside crook of the elbow, or near the ankle joint. In these areas, the rhythmic dilation of an artery can be felt and counted in beats simply by touch (though a stethoscope may also be used).

You can take your own pulse by counting the beats you feel in 30 seconds and multiplying by 2, giving you a beat per minute reading. For example, 40 beats in 30 seconds would be read as 80 bpm. The thumb can have its own pulse rate, so use your index and middle fingers to feel for a pulse.

When taking a pulse, health care professionals measure not only the heart rate, but also the heart's rhythm and the pulse strength. This provides a good indication of your overall heart health and how well it is pumping blood throughout your body. Firm or irregular beats may indicate a problem, and should be discussed with your doctor.

An adult pulse rate between 60 to 80 beats per minute is considered normal, though it can fluctuate as a result of age, exercise, injury, illness, and stress. Pulse rates can also fluctuate greatly

Chapter 4: Vital Signs

within an individual—lower during sleep and higher during periods of strenuous exercise, when more oxygen is required throughout the body. Pulse rates are generally higher in infants and young children and lower in adult athletes. For example, a resting pulse rate in a newborn may measure between 100 and 160 bpm, while an adult athlete may measure only 40 to 60 bpm.

Respiratory Rate

Your respiratory rate is the number of breaths you take per minute—also abbreviated as bpm. It is calculated by counting the number of times the chest expands over a 60 second interval. Observing the respiratory rate and pattern (intermittent periods of hyperventilation, for example) is important when dealing with many respiratory, cardiac, and neurological diseases. Some medications, illnesses, and stress factors may increase the rate of respiration in otherwise healthy people, so it's important for your doctor to examine you for breathing difficulties before measuring respiration. As you've probably noticed after walking briskly or going up several flights of stairs, physical activity can also increase your respiratory rate. Your doctor or nurse will want to measure your rate while you are in a relaxed state.

Infants, children, and teenagers take more breaths per minute than healthy adults. The normal range for respiration in a relaxed adult is between 15 to 20 breaths per minute, whereas babies younger than six months may take 30 to 60 breaths per minute.

Body Temperature

Normal body temperature is in constant flux, and is affected by variables including exercise, eating or drinking, time of day, and a woman's menstrual cycle. It can range between 97.8 and 99.0 degrees Fahrenheit (36.5 to 37.2 degrees Celsius). A temperature above these ranges indicates fever, a sign that the body is addressing abnormal conditions such as sickness or infection.

Body temperature can be measured in several ways:

- Orally, using a glass or digital thermometer. This is the most common method for adults and older children.
- Rectally, using a glass or digital thermometer. Rectal temperatures tend to be .5 to 1.0 degree Fahrenheit higher than oral temperatures.
- By ear. Like rectal temperatures, ear temperatures give higher readings than oral ones. Ear thermometers are popular among parents of young children because they register temperature faster than oral and rectal thermometers, which typically take three minutes to work.
- Beneath the arm, using a glass or digital thermometer. These readings tend to be half a degree Fahrenheit lower than those taken orally.

What Your Signs May Indicate

	High	**Low**
Blood Pressure	Hypertension Stress Pain Response to dehydration	Narcotic use Shock Dehydration
Pulse Rate	Hypertension Stress Pain Response to dehydration	Medication side-effect Heart block
Respiratory Rate	Cold Flu Pulmonary (lung) infection Asthma Chronic obstructive pulmonary disease (COPD) Pulmonary embolism Anxiety Head trauma	Metabolic disorder Tumor Narcotic use Seizure Premature birth
Temperature	Infection Virus Illness Heat exhaustion	Hypothermia Metabolic disorder

CHAPTER 5: RADIOLOGY AND IMAGING

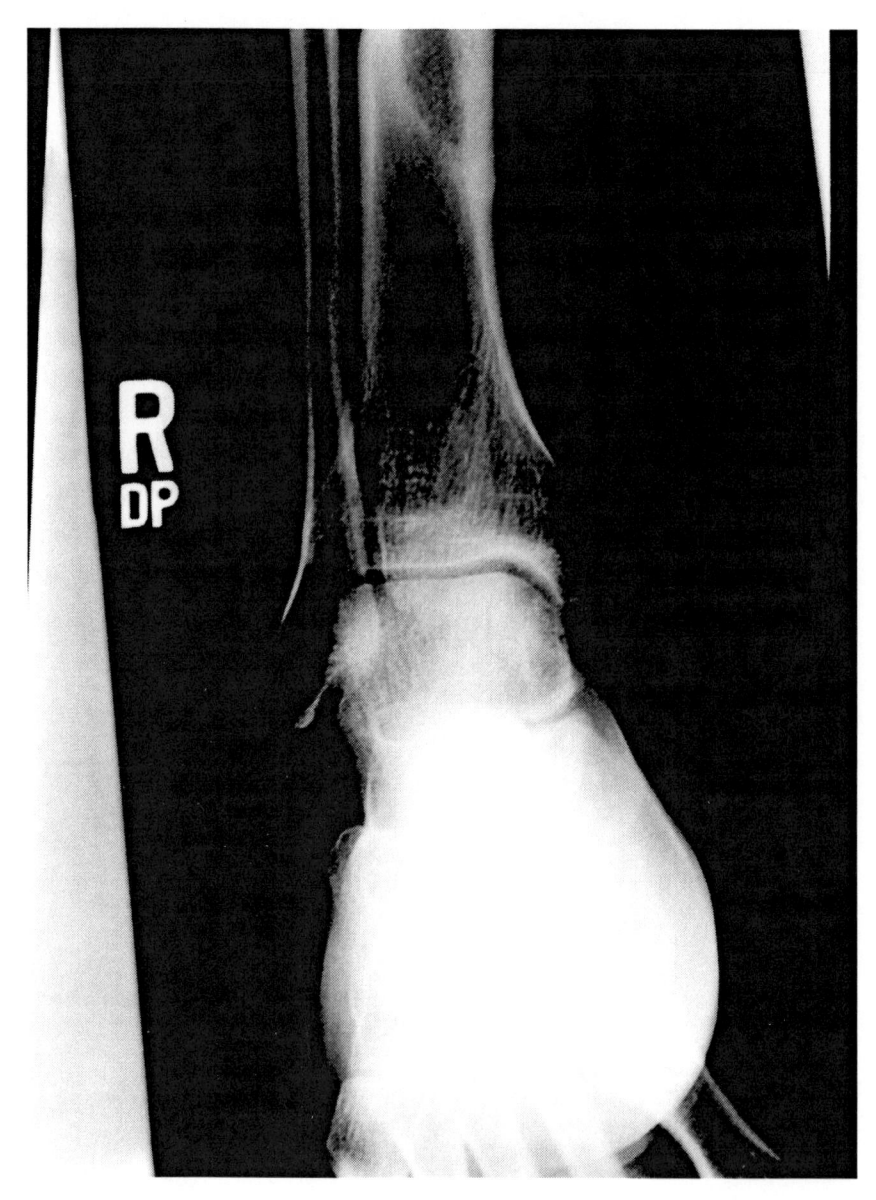

"Oh my God! ... It makes me somehow feel that I am looking at my own death!"

— Anna Bertha Roentgen, wife of the inventor of the X-ray, upon first seeing her husband's creation in 1895

5

RADIOLOGY AND IMAGING

> TOPICS: Arteriogram, CT Scan, MRI, Mammography, Nuclear Medicine, PET Scan, Ultrasound, X-Ray

Today, medical imaging such as X-rays, CT scans, and MRIs are so commonplace that we often forget that the underlying technology is so astounding. These hi-tech "cameras" continue to advance, helping doctors improve their diagnostic abilities and saving hundreds of thousands of lives each year.

Medical imaging has grown in popularity because it provides doctors with accurate two-, three-, and even four-dimensional pictures of everything that exists beneath the skin, including bones, organs, and surrounding internal structures. Using diverse procedures, doctors are now able to detect hidden diseases and pinpoint the slightest abnormalities in the human body at a glance. This allows them to diagnose with greater accuracy and identify potentially harmful issues earlier. All of this is done non-invasively (without cutting skin or otherwise entering the body) through

CHAPTER 5: RADIOLOGY AND IMAGING

the use of external radiation. The three most common categories of imaging are transmission, reflection, and emission imaging, as explained below.

Transmission Imaging passes a beam of light (high-energy photons, to be exact) through the body. The speed of the beam depends on the density of what it encounters. Less dense tissue such as blood, fat, and water appear dark on film, while muscles, ligaments, tendons, and cartilage appear gray, and higher density bones appear white. X-rays, CT (computed tomography) scans, and fluoroscopy are examples of transmission imaging.

Reflection Imaging, such as an ultrasound, sends high-frequency sound waves through the area to be observed. Again, depending on the density of tissue encountered, the speed of the sound waves will vary. A computer then analyzes the data collected and creates the required image.

Emission Imaging uses a scanner to observe nuclear particles or magnetic energy within the body, while a computer assembles this information into a representation of the body section to be examined. For example, magnetic resonance imaging (MRI) uses a large magnet to force changes in body tissue in order to observe magnetic energy in the area of interest.

Though occasionally a patient may need an intravenous line inserted, diagnostic radiology is most often a non-invasive procedure, requiring no cutting or internal equipment. Some minimally invasive surgeries such as arthroscopy ("arthro" = joint, "scopy" = look) use a tiny surgical camera placed inside the body to enhance perspective and improve accuracy in the operating room.

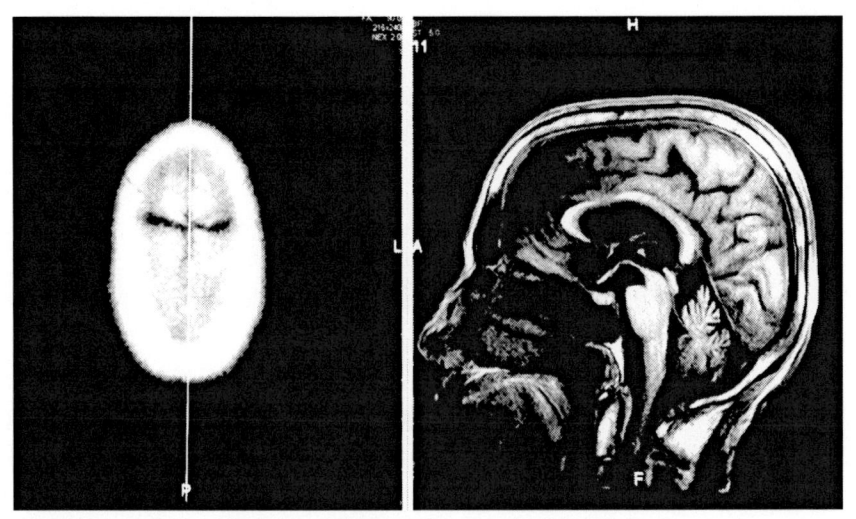

MRI scans such as these are often used to see inside the brain.

Types of Imaging Technology

As medical diagnostic imaging continues to advance rapidly, radiology has become a highly specialized discipline. There is a test for practically every ailment, often requiring one specialist to administer the test, another to read the results, and a third (your doctor) to review results with you. Depending upon your condition, your doctor may order one or more of the following procedures:

X-Rays

X-rays, also known as projectional radiographs, are the most common form of diagnostic imaging, and are used to assess bone injuries, diagnose tumors and dental cavities, and even confirm the existence of pneumonia and other lung illnesses. X-rays use external radiation to pass through the body, leaving a diagnostic image on a specially treated plate or film. From this, a negative is produced as different structures allow for varying degrees of beams to pass through. Softer, less dense tissues such as fat, skin, muscle, and blood allow most of the beams to pass through and appear darker on film. Dense objects such as a bone or a tumor do

CHAPTER 5: RADIOLOGY AND IMAGING

X-RAYS ARE FOR DIAGNOSING
Bone injuries • Tumors • Dental cavities • Lung illnesses such as pneumonia • Problems in the stomach or intestines

WHAT TO EXPECT
1. You must remove any clothing or jewelry that may interfere with imaging or exposure to the body. Gowns are provided if necessary.
2. You sit on or stand beside an X-ray table, positioned between the X-ray machine and film. Other body parts may be covered by a lead apron to reduce unnecessary radiation exposure.
3. The X-ray beam is then focused on the area of interest, and you lie still so as not to blur the picture. The technician leaves the immediate area, and the image is taken.

Often, a series of X-rays is taken from various angles to gain a better perspective of the overall area. It takes just a few seconds for each X-ray to be taken.

not allow the X-ray to pass, so they appear white on the negative. If a bone is broken, the X-ray presents the break as a line across the bone, where the beams were able to travel unrestricted between the two denser pieces.

X-rays are also useful in detecting pathological changes in the lungs and, when used with a radio-opaque contrast material, may be used to visualize the structure of the stomach and intestines.

If you have gastrointestinal issues such as abdominal pain, intestinal bleeding, or changes in bowel habits, your doctor may order a barium X-ray. You will be given a barium contrast agent (by a liquid [oral] dose, IV injection, or, less commonly, anal injection). Barium is a metallic chemical that cannot be penetrated by X-rays. Also called an upper and lower GI series, barium X-rays are essential in diagnosing abnormalities of the gastrointestinal

> **ARTERIOGRAMS ARE FOR DIAGNOSING**
> Artery disease, including arteries in the heart, brain, and kidneys
>
> **WHAT TO EXPECT**
> 1. You sit on an X-ray table and an intravenous line is begun, usually in the arm. An EKG monitor is set to observe the heart and record electrical activity during the procedure.
> 2. A small catheter is inserted into your arm or groin and then threaded into the desired vein or artery.
> 3. You are injected with a contrast material. With the catheter in place, a series of X-rays is taken.
> 4. The catheter is removed and the X-rays are developed for inspection.
>
> This entire procedure usually takes less than an hour.

tract such as tumors, ulcers, and related inflammatory conditions. Complementary procedures such as endoscopic examinations, CTs, MRIs, and ultrasounds are often performed alongside barium X-rays to give your doctors more information on which to base a diagnosis.

Arteriograms

Also called an angiogram, an arteriogram provides an X-ray image of the blood vessels of the heart and vascular system to assess vascular health. (The vascular, or circulatory, system is comprised of vessels that circulate blood.)

This procedure primarily focuses on the ballooning of a blood vessel (aneurysm), narrowing of a blood vessel (stenosis), or potential blockage. Arteriograms rely on the presence of contrast dye that is injected into a patient intra-arterially or intravenously, which causes the blood vessels to appear opaque on the scan, thus enabling doctors to better visualize the health of the ves-

CHAPTER 5: RADIOLOGY AND IMAGING

sels in question. Because arteriograms usually require an arterial puncture, which comes with a high risk of complications and may require an overnight hospital stay, your doctor may recommend a CT scan instead.

Computed Tomography (CT) Scans

Also known as a computed axial tomography (CAT) scan, this technique uses a special camera that rotates around your body to take a series of cross-sectional images called "slices." These slices are then pieced together to create a detailed 3-D image of the inside of an object such as your head, heart, or abdominal cavity.

More detailed than general X-rays, CT scans allow doctors a better view of the organs and internal structures. They are instrumental in diagnosing tumors, determining the source and scope of internal bleeding, and investigating other internal issues. Sometimes contrast substance is injected, taken orally, or less commonly, administered through the anus to enhance the image of the organ or area under study.

Recent advances in CT technology allow cardiologists to quickly view your heart and entire coronary tree (take a deep breath, hold, release, and you're done!), providing excellent resolution and high accuracy in just a matter of seconds. Today some hospitals have scanners that give a 64-slice reading—a vast improvement over the less detailed 4-slice CT scanners that were introduced in the early 1990s.

The 64-slice CT is so fast that its software can "take pictures" of the artery when the dye is in the arterial phase of circulation, which gives this procedure an advantage over an arteriogram. This means contrast can be injected into a vein and still allow a clear picture of an artery. There are rarely complications with CT scans, and you can generally get off of the radiology table and go home. One of the drawbacks of the CT scan is that it delivers more radiation than other imaging diagnostic procedures. Although the

CT SCANS ARE FOR DETECTING
Tumors and other lesions in the chest and its organs • Injuries • Infections • Chest pain • Obstructions

WHAT TO EXPECT
1. All jewelry and clothing that may interfere must be removed. If contrast is to be used, an IV is started.
2. A mild sedative may be prescribed to help you relax during the procedure, which usually takes 30 to 60 minutes.
3. You are placed on a scan table that slides into a large, circular opening of the scanning machine. You are in constant contact with the CT staff, who watch through a window and communicate through speakers. You're also given a call bell to notify staff of any emergency.
4. Once the procedure begins, the scanner rotates around you, emitting low dosage X-rays that are absorbed by the body's tissues, detected by the scanner, and transmitted to the computer. The computer then analyzes this data and creates a corresponding image.

In a few days, possibly sooner, a radiologist will interpret the image and share his or her conclusions with your doctor, who in turn should discuss the information with you.

64-slice CT is a powerful diagnostic tool, its technology continues to be refined, and prototypes for 128- and even 256-slice scanners are currently under development.

Magnetic Resonance Imaging (MRI)
An MRI uses magnetic energy and radio waves to create "slices" or cross-sections of the body for the doctor to view. They produce detailed images by evaluating the activity of hydrogen atoms that have been knocked out of alignment by a powerful magnet. The

Chapter 5: Radiology and Imaging

MRIs ARE FOR DETECTING
Tumors and other lesions in the chest and its organs • Injuries • Infections • Chest pain • Obstructions

WHAT TO EXPECT
1. Due to the strong magnetic field to be used, you will be asked to remove all jewelry and metal objects, including hair pins, glasses, dental pieces, and hearing aids.
2. If you require a sedative and/or contrast medication, an IV is started or medication is taken orally.
3. If you feel uneasy about going inside the confines of the scanner, most labs provide a sleep mask to cover your eyes and headphones to block out the clicking sounds created by the magnetic field and radio waves.
4. You are placed on a table that then slides into the scanner.
5. From the adjacent control room, the MRI staff watches through a window and stays in constant contact through speakers inside the scanner. A call bell is also provided to alert staff if a problem should arise.

You will need to keep very still throughout the procedure, and may be instructed to hold your breath for a few moments.

MRI is an extremely versatile procedure and may be used to examine the brain, heart, internal organs, reproductive organs, and other soft tissue. It is also useful for evaluating blood flow through the body, assessing infections and injuries to the bones and joints, detecting tumors, and diagnosing many forms of cancer.

Performed for in- and outpatient diagnostics, an MRI uses no radiation, unlike CT scans and X-rays. As a result, repeated MRIs generally pose no danger to a patient.

During an MRI, a patient is placed within a large, cylindrical machine. If you are one of the estimated 5 percent of the world's

HOSPITAL STAY

> **ULTRASOUNDS ARE USED FOR**
> Detecting body abnormalities • Treating tumors • Prenatal care
>
> **WHAT TO EXPECT**
> 1. A conducting gel is applied to the area to be evaluated.
> 2. A transducer is used to send ultrasound waves through your body. These sound waves are reflected off structures within your body. This information is analyzed by a computer.
> 3. From this analysis, a picture is created on a television screen and possibly recorded. Your physician uses this picture to assess the area in question.

population with severe claustrophobia, you may be referred to a machine that is "open."

Ultrasound

Medical ultrasonography uses high-frequency broadband sound waves that are reflected by human tissue to produce images. It may be used to evaluate the vascular system, abdominal organs, a fetus, the heart and its valves, the thyroid and related glands, and the urinary tract, as well as joints, tendons, and muscles. This test is used frequently in the ER because it is portable and gives fast, accurate information.

While this form of imaging does not provide as much detail as a CT or MRI, it does allow your doctor to study the function of organs and assess blood flow throughout the body in real time. Ultrasound is also a quick, inexpensive, and safe procedure, emitting no ionizing radiation.

Mammography

A mammogram is both a diagnostic and screening imaging procedure that uses low-dose X-rays to examine the human breast for abnormalities. Mammograms play an important role in the early

CHAPTER 5: RADIOLOGY AND IMAGING

MAMMOGRAMS ARE FOR DETECTING
Abnormalities in the breast

WHAT TO EXPECT
1. You will be asked to undress from the waist up and given a gown. Your doctor will discuss any concerns you may have.
2. The mammography unit is introduced, and your breast is placed between two plates. Pressure is then applied to compress the tissue. This is necessary to produce the best image while using the least amount of radiation. It may cause temporary discomfort.
3. The technician leaves the immediate area. You will then be asked to hold your breath, and the X-rays will be taken. Each breast is X-rayed at least twice from both the top and side.
4. The radiologist reviews the films to determine if additional X-rays are needed and relays this information to your doctor.

detection of breast cancer and serve as a follow up to breast self-examinations. This procedure is believed to be directly responsible for the decrease in mortality rates from breast cancer over the past 30 years.

A standard mammogram usually involves two X-rays of each breast, and is capable of detecting a tumor that cannot be directly felt. Modern advances in mammography have resulted in equipment that provides high-quality images using a very low dose of radiation, and the risks associated with this procedure are considered negligible.

Nuclear Medicine
Nuclear medicine is a specialized branch of radiology that allows doctors to assess organ function as well as structure. It may be used to both diagnose and treat disease.

NUCLEAR MEDICINE IS FOR
Testing organ function • Diagnosing and treating infections and diseases • Monitoring blood circulation

WHAT TO EXPECT
1. A technologist will inject a small amount of a radioactive substance into your body. You may experience some discomfort or pain during the injection. Occasionally, you can swallow the radiation dose in capsule or liquid form, depending on the treatment and organ being studied.
2. The radioactive substance emits "gamma radiation," which is detected by a special camera called a "gamma camera."
3. Once the substance is absorbed (this can take an hour or more), the technologist positions the camera close to the body area being studied.
4. When the radiation dose reaches its destination, the internal image can be displayed on a monitor or placed on film. The pictures should only take a few minutes to complete.

Patients are first given a radioactive material either orally or by injection. Once the material is absorbed by an organ or tissue, images are taken, and doctors can assess the organ's overall health and ability to function. Nuclear scans can be done on the heart, brain, bones, kidneys, thyroid, and breast. They can also be used to monitor blood circulation and diagnose infectious and/or inflammatory diseases. Thanks to their ability to visualize both the structure and functionality of organs and tissue, nuclear imaging procedures are often used to aid in the diagnosis and treatment of abnormalities very early in the progression of a disease, when treatment is often most successful.

Positron Emission Tomography (PET)
PET, pronounced like the word "pet," is a nuclear medicine imag-

CHAPTER 5: RADIOLOGY AND IMAGING

PET SCANS ARE FOR
Detecting diseases and disorders of the heart and brain

WHAT TO EXPECT
1. All jewelry and clothing that may interfere must be removed.
2. A technologist will inject you with a small amount of radioactive material (most commonly fluorodeoxyglucose, or FDG, a sugar). There's usually about a one-hour waiting period between the injection and the actual imaging.
3. Once the substance is absorbed, you are placed on a scan table that slides into the center of a machine.
4. Like a CT, the PET scanner rotates around your body, emitting low-dosage radiation for short periods of time.
5. The time to complete the images may take between 20 and 90 minutes, depending on your size, the area of focus, and the sophistication of the diagnostic equipment in use.

ing technology used to detect cancer, neurological disorders, or cardiovascular disease (i.e., blockages near and around the heart). PET scans produce 3-D images of the body "in action," so doctors can see the size and shape of an organ as well as how well it is functioning. They are often used in conjunction with CT scans and MRIs. Modern PET scanners are equipped with high-end CT scanners, allowing both procedures to be performed during the same session.

After a patient is injected with a small amount of radioactive material, the PET scanner creates an array of images around the patient and detects the signals that are emitted by the injected material. PET records the amount of metabolic activity, and a computer assembles these signals into computer images or films.

Benefits Versus Risks
The evolution of medical imaging into a highly technical diagnostic

field has saved countless lives. Today, doctors are able to make earlier, more accurate diagnoses, and patients often feel more secure knowing they are undergoing the most cutting edge treatment. Still, there are those within the medical community who believe that the benefits of these procedures have led to over-utilization, often in cases where there is little need for the examinations, thus exposing patients to unnecessary radiation, which may increase the risk of cancer.

Since each test is considered separately and any radiation-induced cancer may not appear for years, the true risk posed by a single procedure is difficult to predict. There is no government- or industry-related regulation over the frequency of these techniques. For the sake of a patient's safety, however, most medical professionals agree that CT scans should use the smallest necessary dose of radiation or be replaced by tests such as MRIs or ultrasounds, which emit no radiation. Some estimates conclude that these changes would reduce a patient's exposure by anywhere from 20 to 70 percent.

If you have questions about tests recommended for you, talk to your doctor. It's your doctor's duty to carefully weigh the benefits of any imaging procedure against possible risks, and to ensure you are fully informed.

CHAPTER 6: THE ART OF DIAGNOSTIC MEDICINE

"Medicine is not only a science; it is also an art. It does not consist of compounding pills and plasters; it deals with the very processes of life, which must be understood before they may be guided."

– Paracelsus, Swiss physician, botanist, and alchemist

# HOSPITAL STAY

6

THE ART OF DIAGNOSTIC MEDICINE

TOPICS: Communicating, Doctor-Patient Relationship, Hippocrates, Physical Exam, Questions to Ask, Self-Diagnosis

The potential and performance of health care have changed drastically in the 2,500 years since Hippocrates first referred to medicine as an "art" in his famous oath. Today, doctors continue to pledge to practice medicine ethically and place the patient above all else, as Hippocrates counseled. Still, art can evolve, depending upon the intentions and perspectives of the artist. Medicine is no different.

Thanks to new procedures, scans, and tests, doctors today have greater access to the human body and a fuller understanding of how it works. As this knowledge has expanded, the practice of medicine has grown increasingly specialized and reliant on technology, leaving many patients feeling helpless and confused. At the same time, the Internet now supplies patients with a vast amount of information, some of it valid, some of it not. These trends,

Chapter 6: The Art of Diagnostic Medicine

combined with fear of medical malpractice lawsuits, are straining and reshaping the doctor-patient relationship. Many doctors now tend to forego the more personal physical exam in favor of diagnostic tests, leaving patients feeling left out of the process. As a result, patients may second-guess their doctors, choosing instead to trust an inconsistent network of medical websites while attempting self-diagnosis.

The complex nature of medicine and technology make it increasingly important for medical practitioners and patients to work together. A well-rounded doctor understands the need to personalize the medical experience, incorporating scientific resources with a renewed emphasis on the physical exam and patient history. At the same time, a wise patient must keep in mind that while the Internet may be a practical educational tool, it often provides inaccurate and unverified information. It is no substitute for a medical degree and in-field experience.

The Lost Art of the Physical Exam
Have you ever heard someone complain, "I went for a physical and my doctor didn't even examine me?" For whatever reason, today's patients increasingly see health care as an impersonal scientific study where they are looked upon not as people but as objects to be fixed, a far cry from the days of the house call by a neighborhood doctor. The interpersonal aspect of health care, so essential to a patient's psychological well-being, seems in some respects to have been replaced by a cold, clinical demeanor. Over time, this change has resulted in a rift between many medical professionals and their patients, even as expectations and demands have increased.

Thousands of years ago, medicine offered little more than diagnosis and prognosis. The role of doctor dealt primarily with addressing symptoms and alleviating pain where possible. Physicians predicted the chances of recovery and watched over patients as disease or injury ran its course. Over time, advances in medical technology and the understanding of human anatomy have revolutionized doctors' ability to treat illnesses and greatly altered their responsibilities.

The Rise of Technology

As doctors learned more about the interplay between each of the body's organs, scientific discoveries began to shed new light on the mysteries hidden beneath the skin. The physical exam from the nineteenth century gave rise to the X-ray in the early twentieth century. CT technology in the 1970s was followed by magnetic resonance imaging (MRI) in the 1990s.

Today, the stalwart companion to these radiological tools, the simple blood test, grants further insight into the health of an individual and helps doctors make more accurate diagnoses pertaining to an ever-growing list of diseases. These technological breakthroughs are the reason so many diseases once thought to be death sentences are now often considered routine and treatable.

For example, medical imaging is an invaluable tool for making a quick and accurate diagnosis of a pulmonary embolism, a deadly blockage in an artery of the lung, especially since the accompanying chest pain and shortness of breath could be caused by a multitude of other, less dangerous conditions. Likewise, aortic dissection, a tear in the wall of the aorta, can also lead to unexpected and immediate death if not identified by a physician and confirmed by an angiogram or high speed CT scan with angiographic capabilities.

Without discounting the impact certain advances in technology have had in eliminating or identifying disease, these new tests and procedures were initially intended to augment the information gathered by a physical examination and patient interview. Instead, too often they have become the central focus for many medical practitioners who favor them over the bounty of information gleaned from the appearance and functioning of a patient's eyes, skin, teeth, hair, and reflexes. In the world of modern

CHAPTER 6: THE ART OF DIAGNOSTIC MEDICINE

medicine, ailments are dissected, sectioned off, and divvied out to the appropriate specialist. It can sometimes appear to a hospital patient that the body is no longer seen as a single entity, but rather a collection of parts to address on an individual basis. The art of the physical examination, once so essential to both the diagnostic process and the emotional well-being of the patient, is in danger of losing its integrity.

Back to the Basics

Still, the expressiveness of the human body in times of duress is not completely overlooked. The patient examination continues to lay the foundation for medical students, who must memorize thousands of acronyms and "clinical pearls" to aid in the diagnosis of an ever-growing list of diseases. These mnemonic devices provide an enlightening glimpse at the potency of a well-rounded doctor's powers of observation. For example, when based on blood test results alone, gall bladder disease can be misdiagnosed as appendicitis, resulting in an unusually high incidence of unnecessary surgical procedures. Yet, the time-tested clinical observational pearls of gall bladder disease, the five F's (female, fleshy, forty, fertile and fair), are to this day surprisingly accurate and should never be dismissed if inconsistent with the results of a single diagnostic test.

Of course, a doctor can only be as thorough as time allows, and as medicine shifts toward more lab-oriented methods, it is not only the patient who suffers. With less time to spend getting to know each patient and his or her history, doctors are finding the art of the physical exam more difficult to master. "In 1980 the average length of stay in a U.S. hospital was more than a week," according to Lisa Sanders, M.D., in *Every Patient Tells a Story* (Broadway, 2009). "In 2004, that had dropped to just over three days. So there is less opportunity to do bedside teaching—a triumph of medical economy that only slowly has been recognized to have come at the expense of education. Patients zip in and out of the hospital too quickly for residents to learn from their exams."

This decrease in the practice of hands-on study further emphasizes a purely clinical approach, as contemporary doctors find themselves more and more reliant on lab results. Here are a few ways in which you can improve your doctor-patient relationship and get more out of your appointment:

Questions to Ask Your Doctor
If your doctor prescribes tests, consider asking the following questions. This can help you understand your situation:

> What is the name of the test?
>
> How is the test administered?
>
> What does your doctor expect the test to be able to tell him/her? (Why do you need it?)
>
> Will the test change the outcome of your treatment?
>
> How should you prepare for the test? (Fasting, time of day, exercise?)
>
> Are there risks associated with the test?

The more you know about your doctor's diagnostic approach, the more you will feel like a participant in the process and the more satisfied you will be with the health care you receive. Knowing what to expect plays a significant role in patient satisfaction.

Choosing Your Doctor
For many patients, the psychological aspect of illness is just as

CHAPTER 6: THE ART OF DIAGNOSTIC MEDICINE

> **The Dangers of Self-Diagnosis**
>
> Accurate or not, the Internet makes vast amounts of medical knowledge accessible to patients, so patients often research both disease and cure on their own. Many believe a cure is out there—if only they search hard enough. This can lead patients to expect more from their doctors, viewing them as scientists capable of curing any ailment with a pill or quick procedure, no matter how grave the condition. Such expectations can strain the bond between doctor and patient, as doctors grow tired of being second-guessed by laymen and patients cling to potentially irrelevant facts and unrealistic hopes.
>
> While it's important to take an active role in one's own health care, self-study over the Internet often gives an inaccurate depiction of an illness, and it can easily frighten or send you in pursuit of pipe dreams and panaceas. Remember, some of what you read online may not be general medical consensus, and still more may not pertain to your specific situation.

important as the physiological, and most patients would prefer to feel that they are being helped as a person rather than fixed as an object. That's why it's so important to choose a doctor you trust. Establishing a successful rapport with your doctor requires both sides to work together. As a patient, you expect to be treated with courtesy. So does your doctor. Here are some things you can do to strengthen the doctor-patient relationship:

> **Set your expectations.** Decide what you are looking for in a health care provider when choosing a doctor. Some people want a doctor who is relaxed and personable, while others feel more comfortable with the no-nonsense, take-charge type. Remember, your doctor sets the tone for your initial meeting and will direct the course of your health care. The right fit is important.

Ask friends for recommendations. Go with a known entity whenever possible. A recommendation from a friend, co-worker, or relative gives you some confidence that the doctor knows what he or she is doing. Talking with friends can also be an opportunity to ask questions about the doctor's individual style, demeanor, and medical philosophies from someone with first-hand experience.

Look for chemistry. Like it or not, as in any relationship, the doctor-patient collaboration is based largely on personality. Some people click, while others don't. Since the ability to ask questions is of utmost importance, it is imperative that you and your doctor are able to communicate freely and openly. If you do not like your doctor, you'll be less likely to speak with an open mind, and less willing to trust his or her medical opinions.

Keep things professional. While he or she should be courteous and accommodating, it is important to remember that your doctor is not your friend, but a medical practitioner whose job is to treat you. You should feel comfortable asking pertinent questions, voicing concerns, explaining your medical history, and calling with follow-up issues or to check on lab results. Bear in mind, however, that he or she has a practice to run, and is not at your beck and call. When calling the doctor's office, have your questions prepared and keep conversations concise.

Repeat the facts. Doctors are human. Though they have charts to rely on, they see many people over the course of a day and cannot be expected to remember intimate patient details weeks or months later. If you are pregnant, have an outstanding medical condition, or suffer from drug-related

CHAPTER 6: THE ART OF DIAGNOSTIC MEDICINE

allergies, providing your doctor with a quick list at the beginning of your appointment is a great way to reintroduce yourself so you both can then focus on the matter at hand.

How to Better Communicate with Your Doctor
Once you've chosen a doctor, it's important to prepare so that you get the most out of your first visit. You must express yourself clearly while taking in and understanding the issues and instructions your doctor presents to you. Here is a set of guidelines to assist you in communicating effectively:

Be thorough. Though he or she may be busy, your doctor's primary function is to diagnose your illness, which cannot be done without the facts. It is important to let your doctor know of all your medical concerns during your visit. Not only will this save you from having to return for follow-up appointments, but seemingly unrelated symptoms may turn out to be relevant. Your job is to provide the details on how you are feeling. The doctor will sort it all out.

Be honest. Though talking about health issues can be a delicate matter, remember that your doctor is a professional. A diagnosis can only be as strong as the information you provide, and you are potentially harming yourself by hedging the truth. There is no need to feel embarrassed or uncomfortable about discussing body parts and functions, nor should you feel guilty about lifestyle choices such as sexual practices, smoking, drinking or eating too much, lack of exercise, or recreational drug use. Your doctor has seen it all and is not there to judge, but to heal you.

Ask questions. If something about your condition or treatment is unclear, ask your doctor to repeat it or try to put it in simpler terms. A solid understanding of your situation is vital to recovering your health, be it grasping the facts of your illness or familiarizing yourself with the proper method and dosage of medications. Though your doctor may seem busy, he would much rather explain things during your visit than return a late-night phone call.

Bring a list. Your doctor's office can be a stressful place, making it difficult to think under pressure. Make it easy on yourself by writing out a list of questions or concerns you can read through at the beginning of your visit. Such a list might include anything from the consequences of new or recurring symptoms, lifestyle issues or changes, possible procedures, treatment options, details on disease progression, available support groups, and overall outlook. You may also want to write out a list of both prescription and over-the-counter medications you are currently taking, the dosage, when you take them, and how often. The better prepared you are for your visit, the more relaxed you will be when questioned, and the more you will benefit from your doctor's instruction.

Take Advice. Ask your doctor what changes you can make to your daily habits to prevent a recurrence of your illness or to alleviate symptoms. There is only so much a doctor can do in the office, and she will be glad to note that you intend to pair treatment with a healthier lifestyle at home.

Bring a friend. Though this option may make some people uncomfortable, for others it is a tried and true way to reduce the level of stress brought on by a trip to the doctor's office. The added presence of a trusted friend or relative often makes the whole procedure seem less clinical in nature, and will not only help you to be more honest and objective when

CHAPTER 6: THE ART OF DIAGNOSTIC MEDICINE

relating facts to the doctor, but will provide you with someone to discuss or clarify the doctor's answers and recommendations after the visit.

Educate yourself. If you are the type who needs to know all there is about your disease, by all means do some research. For many, a better understanding of one's illness is a powerful way to reestablish a feeling of control over the situation. Just remember that spending a few hours online does not measure up to a medical degree and hands-on health care experience. Your doctor's opinion as a professional is what ultimately counts. If you do wish to study on your own, ask your doctor what websites, brochures, books, or journals he recommends. If your doctor knows the source of your research, he or she will be in a better position to judge its value and answer your questions honestly.

With all the benefits provided by modern medical technology, the emphasis on diagnostic medicine will continue to influence doctor-patient relationships. It is therefore critical that both sides remember that the practice of medicine is a partnership requiring communication to be successful. Patients must be willing to embrace the technological advances of contemporary health care, while doctors should recognize the positive impact of personal interaction on the well-being of patients.

HOSPITAL STAY

Chapter 7: Surgery Made Simple

"Surgeons must be very careful / When they take the knife! / Underneath their fine incisions / Stirs the Culprit—Life!"

– Emily Dickinson, American poet

7

SURGERY MADE SIMPLE

TOPICS: Appendectomy, Cardiac Surgery, Craniotomy, Invasive and Non-Invasive Surgery, Laparoscopic, Liposuction

If your doctor recommends surgery as a treatment option, you might feel scared or overwhelmed. It's normal to fear what you don't understand. Fortunately, you and your family can rest assured that the procedure you are about to undergo has probably been done many times before. For example, 50 years ago doctors faced huge obstacles when operating on a beating heart, especially since stopping the heart for more than a few minutes increases the risk of brain damage. Today, technology not only makes heart and other forms of surgery possible, many formerly risky procedures are now considered standard.

Each year cardiothoracic (cardio=heart, thorax=chest) surgeons perform more than 500,000 coronary artery bypass grafting procedures (CABG), making this the most common type of heart surgery. Meanwhile, classic (open) appendectomies have been

Chapter 7: Surgery Made Simple

performed by the thousands over the last two centuries—and the number gets even higher when you add the laparoscopic procedures done since 1987. One Texas hospital even reports that it has performed 500 craniotomies (skull/brain surgeries) per year over the last three years.

Invasive, Minimally Invasive, and Non-Invasive

Technology and surgical skill have made advances in many areas. Today, surgeons can stop a heart for hours, perform extensive procedures without risk to tissue or brain function, and remove organs (or unwanted fat) with almost no cutting. Many surgeries can now be performed using measures that are less invasive than ever before. This is why a surgical patient should understand what is referred to as the *invasiveness* of any procedure. For the most part, surgeries break down into three categories: invasive (or open), minimally invasive, and non-invasive.

Invasive (or open) procedures involve making an incision, usually a significant one (in other words, bigger than a tiny cut) in the patient's body, retracting the skin, and inserting instruments or other medical devices. This is what people often think of when they imagine surgery. It's also what has been done throughout history, until the late 1980s.

Many surgical procedures can be done either using minimally invasive or classic open techniques. Which option you receive normally depends on a number of factors—including the ailment being treated as well as your state of health and age.

Minimally invasive (or endoscopic, keyhole, pinhole, or band aid) procedures usually involve tiny incisions and minimal body intrusion. Sometimes they are referred to in terms that indicate the area of the body where the surgery is being performed (for example, laparoscopic refers to surgeries in the abdomen, while thoracoscopic refers to surgeries in the chest cavity).

The procedure could be relatively simple (getting a shot, for instance) or more involved (like endoscopy—used to take images or small amounts of tissue, aka a biopsy, by inserting a small scope into the body via an existing anatomical opening). These surgeries can take longer to perform but often involve shorter hospital stays, and many can even be done on an outpatient basis. Patients frequently experience less pain, scarring and complications post-surgery. But minimally invasive does not necessarily mean minor surgery. Each procedure comes with risks and the chance for potentially life-threatening complications.

Often, remote-controlled equipment such as miniature video cameras and fiber optic cables are used in these procedures, as well as special surgical tools that are threaded through a small incision or orifice in the body. These cameras transmit pictures that allow the surgical team to maneuver the tools as necessary.

Non-invasive procedures do not break the skin, penetrate a body cavity, or remove biological tissue. In other words, there are no incisions. Most of the tests done during an annual physical fall into this category: taking your pulse, monitoring blood pressure, and listening to the heart and lungs. But a series of diagnostic images and signals such as x-rays and ultrasounds are also non-invasive, as are treatments and therapies like radiation therapy, defibrillation (shocking the heart), and biofeedback. Some familiar non-invasive procedures include: magnetic resonance imaging (MRI), positron emission tomography (PET) scans, electrocardiography (EKG), and breath tests (like a breathalyzer).

Non-invasive procedures usually don't scare patients—though some of their machinery used in the process can be a bit intimidating. They are frequently used in conjunction with other procedures to monitor vitals and ensure patient stability.

Common Hospital Surgeries

This chapter explores some common types of surgery—appendectomy, breast mass and ovarian cyst removal, heart surgery,

> **WHAT DOES THE APPENDIX DO?**
> Located in the lower right quadrant of the abdomen, the appendix has the dubious distinction of being discussed only when infected. Some scientists believe that the appendix serves a modern biological purpose, catching and retaining beneficial bacteria helpful to the colon. Most doctors, however, feel that it is left over from a time when the human diet relied far more heavily on foliage.
>
> Whatever its role, the appendix, measuring an average length of 8 to 10 centimeters, forms a wormlike extension at the beginning of the body's large intestine, called the cecum. The cecum serves as a large pouch for receiving waste from the small intestine. When the opening that leads from the appendix to the cecum becomes blocked by mucus or stool, the bacteria residing within becomes trapped and begins to penetrate the appendix wall. In reaction, the body's natural defenses cause the appendix to become inflamed and swollen. At this point, surgical intervention may be necessary, because if the infection spreads through the wall of the appendix it could cause the appendix to rupture and allow access to the abdominal lining and the peritoneal cavity, where vital organs such as the pancreas, stomach, and spleen are located. Traditionally, the appendix is removed through an incision in the right lower abdominal wall.

craniotomy, and liposuction—to help patients understand how invasive, minimally invasive, and non-invasive techniques play a role in surgery today, as well as what may be involved.

Appendectomy
The first report of an appendectomy came in 1735 from a surgeon in the English army who performed the operation without anesthesia. Today, one out of every 2,000 people has an appendectomy, almost always with pain medication.

Although appendicitis is one of the more frequent surgical emergencies, there is no specific test to diagnose it with absolute certainty. Symptoms typically include abdominal pain. During early stages, the pain can be difficult to pinpoint, as inflammations of the small intestine and colon are often localized, but other symptoms may include loss of appetite, fever, and/or nausea.

Because many conditions exhibit similar symptoms (such as kidney disease, pancreatitis, right sided diverticulitis, and pelvic inflammatory disease), hospital staff must rely upon their expertise to examine the results of diagnostic tests while observing the patient and deciding whether to fight the inflammation with antibiotics or recommend surgery for verification and/or treatment.

Along with a physical and complete patient history, a doctor may order several diagnostic tests if appendicitis is suspected. These include a complete blood count (CBC), urinalysis, and if necessary, radiology studies such as a CT scan, ultrasound, or an abdominal X-ray.

The current standard for treating typical appendicitis is an appendectomy, either open or laparoscopic.

Laparoscopic Appendectomy

The laparoscopic technique involves making three or four tiny incisions and inserting a miniature camera and surgical instruments into the abdomen. The camera relays an image of the surgical area on to a monitor in the operating room, allowing the surgeons to guide the instruments and remove the appendix. This option can treat most cases of acute appendicitis. The benefits include:

- Less post-operative pain
- Faster recovery and return to normal activity
- Shorter hospital stay
- Fewer post-operative complications
- Minimally sized incisions/scars

CHAPTER 7: SURGERY MADE SIMPLE

In most cases, patients are discharged within 24 to 36 hours. But some patients are not appropriate candidates for laparoscopic surgery, usually due to certain pre-existing conditions, weight issues, previous abdominal surgery, or an extensive infection.

After the operation your body will need time to heal. While getting out of bed the day after surgery can help reduce the risk of blood clots and sore leg muscles, don't expect to return to activities like walking up stairs or driving for a week or two.

Open Appendectomy
Though an invasive (open) procedure is more involved, it's easy to understand the nine basic steps surgeons typically follow:

1. Make an incision in the skin and adipose tissue (protective padding just under the skin and around the organ) and retract with a special surgical instrument.
2. Clamp, cut, and tie the appendix and nearby blood vessels.
3. Dissect through the adipose layer, grasp the fascia (connective body tissue) with two clamps, and then make another incision using an electric cautery knife or scalpel.
4. Dissect through the muscle in the same way, extending the incision.
5. Grasp the peritoneum (thin membrane that lines the abdominal and pelvic cavities) with two clamps, incise with the deep knife, and again extend the incision.
6. Grasp the appendix with a clamp. Cultures are sometimes taken at this time to identify the type of bacteria, especially if the appendix has ruptured.
7. Clamp the base of the appendix and place an O-Chromic tie (which releases on its own in about 3 weeks) at the end.
8. Remove the appendix. Place a circular suture—called a purse string suture—through the external layer of the "stump" of the appendix and extend it around its circumference.

Complete the suture by pulling the two ends together tight, like a purse string, then tie the "strings" and cut.
9. Close the peritoneum with suture of choice. Drains may be used for wounds that are not clean (i.e., ruptured appendix), but are generally not needed.

After an open appendectomy operation, you can expect to recover in the hospital for two to five days, barring any complications.

Breast Mass and Surgery
You've probably seen advertisements recommending adult women regularly screen themselves to detect a breast mass, also called a "lump." The primary concern when dealing with any breast mass is the possibility of cancer. Breast masses are either benign (not harmful or life-threatening) or malignant (possibly life-threatening). Benign breast masses have several causes, including fibrocystic disease, fibroadenoma, intraductal papilloma, and abscess.

If your doctor finds a breast mass, there is no immediate cause for concern. Most masses are benign. Still, a general surgeon must examine any patient with a breast mass, and to render a definitive diagnosis, the surgeon may need to obtain a specimen of the mass for pathological examination. Depending on the size, location, and characteristics of the breast mass (including whether or not it can be felt by touching the patient's skin), your doctor has a few options for the breast biopsy. These include an open excisional biopsy (using local anesthesia, the doctor removes part or all of the mass), an axillary node dissection (performed under general anesthesia, this can determine if the cancer has spread to other parts of the body), a sentinel node dissection (also performed under general anesthesia, this is used to identify and remove lymph nodes that may contain cancer), and a needle aspiration (using local anesthesia, this procedure will investigate a superficial breast mass).

CHAPTER 7: SURGERY MADE SIMPLE

Treating an Ovarian Cyst
Ovarian cysts, small fluid-filled sacs that develop in a woman's ovaries, are another common reason for surgery. Generally harmless, cysts can be problematic due to possible rupturing, bleeding, or pain. Often, the pain associated with an ovarian cyst can be treated with over-the-counter pain relief medication such as ibuprofen (Motrin) or acetaminophen (Tylenol). Sometimes stronger pain medication may be prescribed.

In situations where the cysts are large and do not go away on their own, a doctor may need to surgically remove them. While some surgical procedures successfully remove the cysts without causing any damage to the ovaries, on occasion it may be necessary to remove one or both ovaries completely. Surgical intervention can be done laparoscopically (involving a small incision into the abdomen and a very quick recovery, within days) or with a laparotomy (a more invasive procedure involving a larger incision through the abdominal wall and a longer recovery, within weeks). Both options usually require general anesthesia. In cases where a cyst has twisted (ovarian torsion), causing the patient extreme pain (and often nausea and/or vomiting), an emergency procedure may be required.

Craniotomy (Brain Surgery)
Though brain surgery is considered by many to be an example of a complicated procedure, it is surprisingly straightforward. It's also one of the earliest surgeries to have been performed. Remains dating back to the Neolithic period (circa 7000 B.C.) show evidence of surgical tools and successful brain operations. The belief is that even early man recognized the value of drilling into the skull to relieve unwanted pressure or treat illness.

Today, craniotomies are most often critical surgeries done to address brain lesions or traumatic injuries. They may also be used to implant deep brain stimulators for the treatment of Parkinson's disease or epilepsy. Some life-threatening conditions which may necessitate this procedure include:

- Brain tumors
- Bleeding (hemorrhage) or blood clot (hematoma) from injuries (subdural hematoma or epidural hematoma)
- Weaknesses in blood vessels (cerebral aneurysms)
- Abnormal blood vessels (arteriovenous malformations)
- Damage to tissues covering the brain (dura)
- Pockets of infection in the brain (brain abscesses)
- Severe nerve or facial pain
- Trauma to the skull and reparation of skull fractures
- Some forms of seizure disorders (epilepsy)

Craniotomies are normally preceded by an MRI or other diagnostic imaging procedure designed to give the surgeon a precise map of the patient's brain so he or she can determine the necessary angle of access to the appropriate part of the brain. During a craniotomy, part of the skull, called a "bone flap," is removed to allow the surgeon access to the brain itself. The amount and placement of skull to be removed depends greatly on the patient's condition.

Minimally Invasive Craniotomy

Today, surgeons employ burr holes, sometimes referred to by the term "keyholes," for smaller, dime-sized craniotomies. (Power drills and saws have replaced earlier tools.) Stereotactic frames, image-guided computer technology, or even endoscopes can complement less invasive procedures typically used to:

- Insert a shunt to drain cerebrospinal fluid (hydrocephalus)
- Insert a stimulator to treat Parkinson's disease
- Insert an intracranial pressure (ICP) monitor
- Perform a needle biopsy

CHAPTER 7: SURGERY MADE SIMPLE

- Drain a blood clot
- Remove small tumors and aneurysms

Skull-based surgery usually refers to more complex craniotomies involving the removal of a portion of the skull covering a sensitive part of the brain. Sophisticated computer technology is often necessary to assist surgeons with these more complicated surgeries, especially when removing large brain tumors, treating aneurysms, or repairing the brain following a skull fracture or injury. Surgeons often rely upon highly specialized equipment and techniques, such as image-guided surgery (IGS). This is a stereotactic (three-dimensional coordinate system for locating small targets) technique that, with the assistance of CT or MRI scans, pinpoints the exact location of a lesion by using a lightweight head frame attached to the patient's skull.

Open Craniotomy

An open craniotomy, while still a skilled procedure, entails only a few basic steps:

1. The hair on one section of the scalp is shaved.
2. The scalp is cleansed and prepared for surgery.
3. An incision is made in the scalp and a hole is drilled through the skull.
4. A piece of the skull is removed, usually temporarily.
5. The necessary surgery is performed.

A surgeon typically opens the skull in one of two ways: (1) By making an incision from behind the hairline, in front of the ear, or just above the eye; or (2) By making an incision at the nape of the neck, near the occipital lobe (the bump on the back of your head). The surgeon marks the target area with a felt tip pen and makes an incision up to the thin membrane covering the skull. During this process, it is necessary to seal the numerous small arteries in the scalp so that the surgeon can minimize bleeding. The skin flap is then folded, exposing the skull.

Next, a high-speed hand drill or an automatic craniotome creates a circle of holes in the skull and places a soft metal guide underneath the skull from hole to hole. This guide aids a fine wire saw to cut from under the bone and remove the appropriate part of the skull, exposing the brain. Then the surgery really begins.

After the surgery is complete, the bone is replaced and secured with fine, soft wire. The surgeon then sutures the skin and tissue of the scalp. Most small skull holes heal quickly and without complication.

All craniotomies carry such risks as possible swelling of the brain, infection (a risk in any surgery), and seizures, and you may need oxygen and/or medication to control or minimize these concerns. There may also be drainage from the head, depending on the reason for the craniotomy. Still, on average you will be out of bed within a day, and out of the hospital within a week. Quick recoveries usually follow successful surgeries.

Cardiac Surgery
If you are scheduled for heart surgery, you are in good company. Many political figures and celebrities have entrusted the repair of this vital organ to health care professionals. Some fellow patients you may recognize include photographer Ansel Adams, author Isaac Asimov, basketball coach Red Auerbach, former first lady Barbara Bush, talk show host Johnny Carson, former U.S. President Bill Clinton, businessman Ben Cohen (of Ben & Jerry's), actor Patty Duke, former U.S. Secretary of State Henry Kissinger, singer Peggy Lee, talk show host David Letterman, journalist Bill Moyers, talk show host Regis Philbin, actress Dame Elizabeth Taylor, actor Burt Reynolds, and actor/comedian Robin Williams.

Heart surgery is quite possibly one of the greatest medical

CHAPTER 7: SURGERY MADE SIMPLE

advancements in the past 30 years. Coronary artery bypass grafting (CABG) and replacement and repair of the aortic and mitral valves are among the most frequently performed heart operations in the U.S. today, and an increasing number of these surgeries use minimally invasive surgical techniques rather than the conventional open-chest surgery.

Even the most complex-sounding heart surgeries can be broken down into easy steps. For example, during CABG (or bypass) the surgeon takes a vein or artery from the chest, leg, or other body part and connects it to the blocked artery. This process is called grafting. The grafted artery bypasses the blockage by going around the blocked site. This biological "detour" allows oxygen-rich blood to reach the heart muscle. If necessary, a surgeon can bypass as many as four or five blocked coronary arteries in one surgery, which explains the terms "quadruple" and "quintuple" bypass surgery.

Multiple machines and tools assist in the bypass. The invention of the heart-lung machine, which takes over the work of the stopped heart, and the development of advanced body cooling techniques, which allow for longer surgical times without a beating heart, have provided surgeons with a wider window in which to perform their procedures and address any complications.

A heart-lung machine, also called a cardiopulmonary bypass machine, functions in place of the heart by replacing its pumping action and adding oxygen to the blood. This ensures that the heart will not move during the operation, which is necessary for open-heart surgery. The heart-lung machine can take over the work of the heart and lungs for hours, granting surgeons the time needed to accomplish their goals without fear of causing permanent damage to other organs.

A heart-lung machine works just like the human heart. The machine carries blood from the upper-right chamber of the heart (the right atrium) to a special reservoir called an oxygenator.

HOSPITAL STAY

MOST COMMON REASONS FOR HEART SURGERY
- Repair or replace the valves that control blood flow through the chambers of the heart.
- Bypass or widen blocked or narrowed arteries to the heart.
- Repair aneurysms, or bulges in the aorta, which could be deadly if they should burst.
- Implant devices to regulate heart rhythms.
- Destroy small amounts of tissue that disturb electrical flow through the heart.
- Make channels in the heart muscle to allow blood from a heart chamber to flow directly into the heart muscle.
- Boost the heart's pumping power with muscles taken from the back or abdomen.
- Replace a damaged heart with a heart from a donor.

Inside the oxygenator, oxygen bubbles up through the blood and enters the red blood cells. This causes the blood to turn from dark (oxygen-poor) to bright red (oxygen-rich). A filter then removes air bubbles from the oxygen-rich blood, and this blood travels through a plastic tube to the body's main blood passageway (the aorta). From the aorta, the blood continues to flow throughout the rest of the body.

Minimally Invasive Cardiac Surgery

In a minimally invasive procedure, surgeons access the heart through small incisions in the chest to avoid splitting the breastbone (called a sternotomy). Some hospitals use endoscopic technology, some use newer, robotic technology, and some offer both options. Either way, this minimally invasive procedure can provide surgeons with an extensive view of the heart from more angles than are available with traditional open surgery. As with other types of surgery, this option reduces pain, recovery time, and scarring. But unlike other types of surgery, the heart must be stopped

Chapter 7: Surgery Made Simple

and the flow of blood must be diverted (using cardiopulmonary bypass—via the heart-lung machine).

Open Heart Surgery
Traditional open heart surgery has been performed since World War II, when U.S. Army surgeon Dwight Harken pioneered the removal of shell fragments and bullets lodged inside the heart. By practicing on animals, Dr. Harken developed a technique to cut into the wall of the heart and use a finger to sweep out the shrapnel. The hearts that Dr. Harken worked on were still beating during surgery. Thankfully, this isn't the case anymore. Today, open-heart surgery breaks down as follows:

1. Make an incision (approximately 11 to 12 inches long) in the chest and split the breastbone.
2. Retract the breastbone and ribs to open the chest and access the heart area.
3. Reroute the functions of the heart (blood flow/oxygenation) through a heart-lung machine.
4. Inject a heart-paralyzing solution to stop the heart for the duration of the procedure.
5. After the procedure, close the incision and restart all heart functions.
6. Make sure the heart is beating strongly, and then disconnect the heart-lung machine.
7. Close chest incisions.

Following surgery, your heart and blood pressure should be closely monitored over the next 12 to 24 hours, and medication may be used to manage pain, regulate circulation and blood pressure, and prevent coagulation (thickening) of the blood. The hospital stay is usually four days to a week, but it can sometimes be considerably longer.

Liposuction

Liposuction is an elective surgery (meaning not essential to maintaining health) that almost always uses minimally invasive techniques. Although there has been publicity surrounding less invasive laser-based procedures that attempt to destroy or dissolve the fat, liposuction traditionally involves small, barely noticeable incisions. In fact, even the laser version requires some "cutting" to suction out the freed fat.

By definition, liposuction is the removal of fat deposits using a tube that is inserted beneath the skin. The fat is then sucked out using a vacuum-like device. The procedure is often performed on the abdomen, buttocks, hips, thighs, and upper arms. Recovery times range from one to four weeks. Because swelling may last up to six months, the results of the fat removal may not be visible for up to a year after surgery.

Surgery Isn't That Complicated

Fearing surgery is natural and normal, but learning more about what your procedure involves can make things a little less frightening. Surgeries such as cardiac, brain, breast mass, and liposuction are performed every day around the world. So while this may be your first surgical experience, you can take comfort knowing that it is not the first for your surgical team.

Chapter 8: Anesthesia

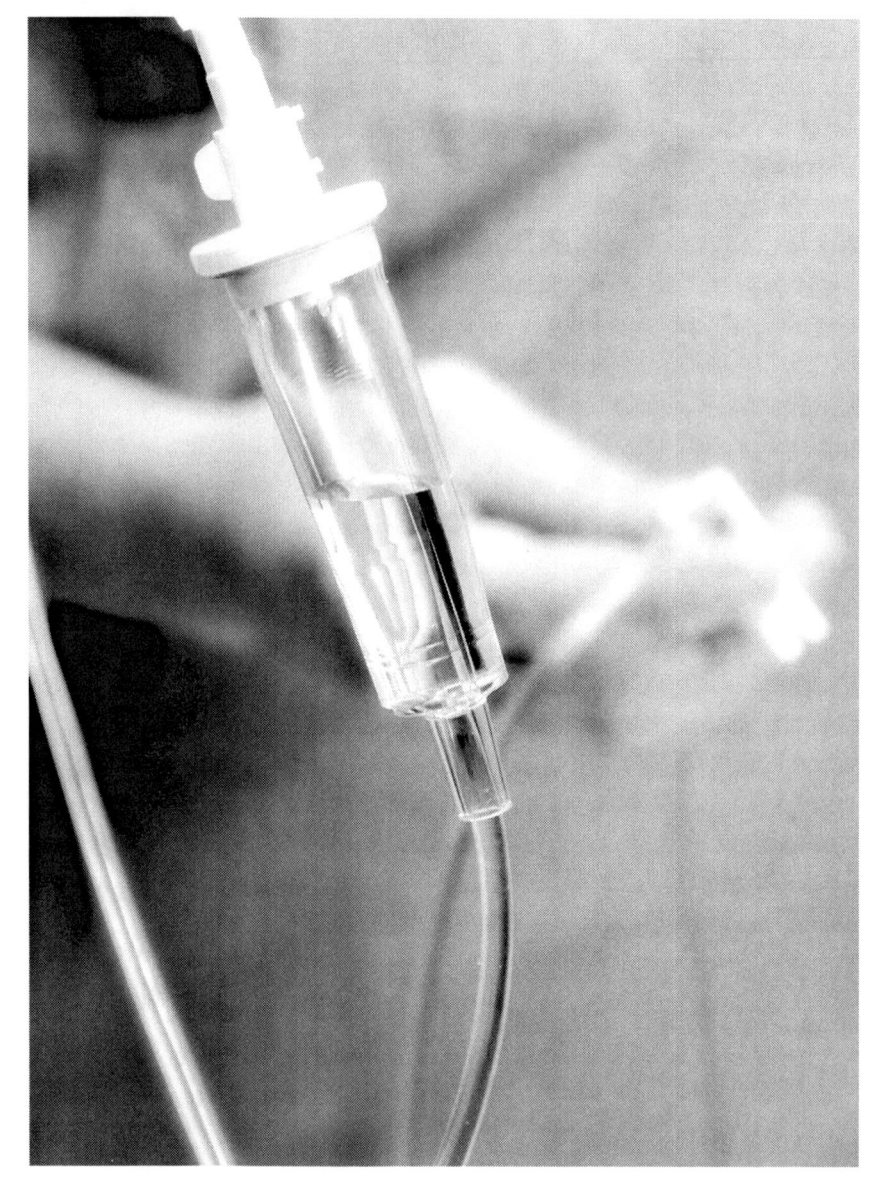

"To be in love is merely to be in a perceptual state of anesthesia."

– H.L. Mencken, American journalist and magazine editor

8

ANESTHESIA

TOPICS: Anesthesiologist, Conscious Sedation, Epidural, Fentanyl, General and Local Anesthesia, Intubation, Novocain

Anesthesia, derived from the Greek word *aisthesis*, meaning "without feeling," refers to the state of being temporarily without sensation or awareness. While that may sound scary, it's actually quite the opposite. Anesthesia enables you to undergo medical procedures without experiencing pain or distress. After all, most of us would not want to be conscious during open heart surgery, watch our own mastectomy, or feel a tooth being drilled.

Anesthesia varies in its effects, depending upon whether the chosen anesthetic works inside or outside the brain. The most common types of anesthesia are general anesthesia, conscious sedation, regional anesthesia, and local anesthesia. General anesthesia renders you unconscious and prevents you from feeling pain during your procedure. Conscious sedation also prevents you from

Chapter 8: Anesthesia

feeling pain, but it enables you to stay drowsy and awake. Regional anesthesia blocks pain in a particular area by targeting specific nerves. Local anesthesia causes you to lose sensation in a small area for a minor procedure. It is the only type of anesthesia that can be administered by a health care worker who is not a licensed anesthesiologist or nurse anesthetist.

Of all the types of anesthesia, general anesthesia requires the most specialized training. Just as the surgeon must work with precision, the anesthesiologist must administer general anesthesia in such a way as to help the patient "sleep" long enough to get through surgery with no recollection.

How Does General Anesthesia Work?

Though today's doctors have a heightened understanding of medications in general, modern science still doesn't know exactly how general anesthetics work. Although the benefits of general anesthesia are accepted by the medical community, no one has yet been able to pinpoint the exact means by which anesthesia achieves its effects.

The most commonly accepted theory is that general anesthesia instructs the central nervous system to temporarily restrict the body's ability to transmit neural impulses between adjacent neurons, effectively plunging you into a deep sleep where no pain is felt and no conscious memory is created. Certain anesthetics may also focus on the spinal cord, relaxing the body's muscles and reducing unnecessary movement.

Several factors influence the way you respond to anesthesia. General anesthesia is most often administered in "cocktail" form, meaning as a mixture of medicines—not as a single drug, as most people believe. Within this mixture, each drug has its own duration and effects, as well as its own way of working to achieve the required ends. Each possible combination can be modified depending on the other substances with which it is mixed. The way the anesthesia is administered—orally, intravenously, or inhaled as a gas—also affects the way it works.

What Happens to Your Brain During General Anesthesia?
While under the effects of general anesthesia, your body and brain essentially "go to sleep." As a result, you neither feel nor remember your medical procedure. Sleep induced by anesthesia, however, is different from ordinary sleep, as the anesthetized brain does not respond to pain and remembers nothing.

The Evolution of Anesthesia
For thousands of years, people have sought ways to numb pain and dilute consciousness. Dating as far back as 1500 B.C., the earliest anesthetics were herbal opiates. People in India and China experimented with cannabis incense and wine, while classical Greek and Roman medical texts noted the numbing properties of opium. Inhalant and oral anesthetics were first used in Arabia and Persia, while in Islamic Spain, narcotic-soaked sponges were placed over the face to create the same effect.

In the United States, before the mid-1800s, surgical patients were given alcohol or opiates, such as laudanum, and expected to manage the pain as best they could.

Surgeons continued to experiment in the early twentieth century. Morphine was introduced, and ether and chloroform were replaced by halogenated hydrocarbons (such as halothane and isoflurane). In 1942, doctors tested patients using curare (plant-derived poisons) to reduce reflexive responses to surgical incisions, while also providing artificial respiration to assist the patient's breathing. This is now known as "balanced anesthesia."

Modern Anesthesia
Today, patients receiving anesthesia will likely receive a combination of medications. For surgical procedures requiring general

CHAPTER 8: ANESTHESIA

anesthesia, a patient might first be given a sedative (often the medication Versed) intravenously to relieve anxiety, induce sleep, and relax the muscles. A narcotic analgesic (often Fentanyl) may follow, also intravenously. Next, a short-acting hypnotic drug such as Propofol may be given to ensure deep sedation. Once the patient is completely sedated, the anesthesiologist may administer Sevoflurane (in a mixture of nitrous oxide and oxygen) for the duration of the surgical procedure.

While regional and local anesthetics such as epidurals and Novocain (which you may have experienced at the dentist's office) cause loss of feeling in specific parts or areas of the body without altering consciousness, general surgical anesthesia affects the receptors of the brain, desensitizing both body and mind. General anesthesia is usually administered intravenously (through a vein), or occasionally via inhaled medication (such as old-fashioned laughing gas). It can also be administered intramuscularly (through a muscle); rectally (for bowel surgery or to reduce the risk of vomiting); or through the gastrointestinal tract. The result is a complete loss of consciousness during which you cannot be woken or stirred, even with stimulation.

Once unconscious, you may need assistance with breathing, and if so you will be given oxygen. In certain cases, a breathing apparatus (such as a breathing tube) may be used to help you maintain an open airway, a procedure referred to as *intubation*. (Remember that you are asleep for all of this, and you will most likely remember nothing of these events.)

The effects of general anesthesia are beneficial both during and after surgery. During surgery, general anesthesia prevents motor responses, as well as autonomic (nerve) and skeletal (bone) responses to certain stimuli (such as when a surgeon makes his or her incision). In other words, all systems of your body will remain still so the surgeons can do their work. After surgery, you should have no recollection of either the pain inflicted during the procedure or any corresponding events (often called postoperative amnesia). In many ways, it is as though you are "turned off" while surgery is performed. The body feels no pain, while the mind remembers nothing of the experience.

Potential Risks
Though in mainstream use throughout America's hospitals, general anesthesia is a complex procedure with its share of risks. For the most part, your level of risk depends on your age, your overall health, the type of anesthetic, and your body's reaction to the substance once administered. Certain medical conditions such as heart, circulation, brain, liver, or nervous system disorders may increase the risk of complications, even though these factors are usually taken into account when the anesthetic is chosen.

Since general anesthesia can disrupt the normal throat reflexes that prevent aspiration, such as swallowing, gagging, and coughing, an endotracheal tube is often inserted into the windpipe to prevent liquid from entering the respiratory tract. This ensures that the contents of your stomach do not enter your lungs. To further reduce the risk of aspiration, you may be instructed not to eat or drink anything for several hours before surgery.

One potentially serious risk from anesthesia is allergic reaction to the medications used. This is normally seen immediately after the substance has been administered and dealt with via a quick change in medication prior to surgery by the monitoring anesthesiologist. Other serious risks include changes in heart rate, rhythm, or blood pressure, which could result in a heart attack or stroke (though these could also be attributed to the effects of surgery).

With advances in the science of anesthesiology and monitoring technologies, the associated risks have declined significantly over the past 20 years. Most hospitals employ an experienced anesthesiologist whose job is to administer the anesthetic and monitor your body chemistry and functions such as breathing,

CHAPTER 8: ANESTHESIA

heart rate, and blood pressure throughout the surgery. Although patients with existing health conditions can be at greater risk, in healthy patients the danger of serious injury or death as a result of anesthesia is extremely low—less than one in 250,000.

While the processes associated with general anesthesia continue to mystify the modern medical community, their effects open new doors for surgeons and enable the use of new technologies. Today's anesthesiologists may not fully understand the reasons these tools do what they do, but they have learned to control them and guard against misfortune as they lower patients into the depths of unconsciousness. So, if you find yourself headed for the operating table, frightened at the prospect of "going under," realize that the only alternative is staying awake. *That* you would remember.

COMMON TYPES OF ANESTHESIA

General Anesthesia
Renders you unconscious and prevents you from experiencing pain or discomfort during a surgical procedure.
- Used for surgical procedures
- Given via injection, gas, or intravenously

Conscious Sedation
Controls consciousness during a procedure. You remain drowsy and awake, but experience no pain. You'll likely have no recollection of your procedure. Recovery time is much faster than if you undergo general anesthesia.
- Used for painful/uncomfortable procedures such as a colonoscopy
- Given intravenously

Regional Anesthesia
Blocks pain within a specific area by addressing the nerves.
- Used for pain associated with childbirth
- Given intravenously

Local Anesthesia
Causes you to lose sensation in a small area.
- Used for minor procedures such as dental work or stitching
- Given via injection

Chapter 9: Hyperbaric Oxygen Therapy

"What oxygen is to the lungs, such is hope to the meaning of life."

— Emil Brunner, Swiss theologian

9

HYPERBARIC OXYGEN THERAPY

TOPICS: The Bends (Decompression Sickness), Carbon Monoxide Poisoning, Hypoxia, Monoplace and Multiplace Chambers

Hyperbaric oxygen therapy is a medical treatment in which you breathe pure oxygen within a sealed, pressurized chamber. It may sound futuristic, but oxygen therapy has been in use for more than a hundred years. Today, hyperbaric oxygen therapy is a staple in health care facilities around the world, and it is used to treat everything from decompression sickness, cyanide poisoning, and severe carbon monoxide poisoning to skin injuries and infections, radiation poisoning, and brain disease.

Historical Uses

Oxygen therapy in contained chambers dates back to 1662, although it was not used for medicinal purposes until the end of the 19th Century. In 1921, an American doctor named Orville Cunningham noticed that people with certain heart diseases fared

CHAPTER 9: HYPERBARIC OXYGEN THERAPY

better if they lived at sea level rather than in the mountains. He attributed this to the increase in oxygen, and treated heart patients accordingly with positive results.

After World War I, the U.S. military tried to eliminate "narcosis" (an alteration in consciousness that can occur during deep sea diving) by modifying the breathing mixture used in submarines. In the 1930s and 1940s, civilian doctors began using hyperbaric oxygen therapy to treat deep-sea divers suffering from "the bends," or decompression sickness, an extremely painful and sometimes life-threatening condition affecting divers who surface too quickly.

Over the next 20 years, researchers discovered numerous additional uses for oxygen therapy and recognized that a temporary increase in oxygen levels had benefits that went far beyond treatment of the bends.

What to Expect

Modern medicine has taken hyperbaric treatment out of the water and into the hospital with the invention of monoplace (single) and multiplace (multiple) chambers that flood with pure oxygen or compressed air, often using specialized gas masks.

Normally, the procedure consists of a series of one-hour sessions over a period of days, and is usually painless, with the exception of a pressurized sensation in the ears similar to that of being on an airplane. Occasionally, the pitch of the voice may sound like a cartoon character and slight changes in vision may occur, both of which are temporary.

Once the desired pressure is reached, a mask is applied and you are encouraged to read, sleep, or relax. You are in constant contact with the technician through the use of a two-way intercom for the duration of the session.

During treatment, your body's tissues are saturated with oxygen, reversing the ill-effects of hypoxia (low oxygen levels) while assisting in the growth of new blood vessels, deactivation of toxins,

THE IMPORTANCE OF OXYGEN
It is no wonder that powers of healing are ascribed to oxygen. Oxygen is one of Earth's most versatile elements. It makes up one-fifth (21 percent specifically) of our atmosphere, combines with most elements, is necessary for both plant and animal respiration (breathing), and plays a key role in combustion (such as when lighting a campfire). Without it, life as we know it on this planet would cease to exist.

and a decrease in swelling and inflammation throughout the body. This procedure has also been proven to expedite the rate of healing and increase the body's ability to ward off infection. In light of such effectiveness, oxygen therapy has grown in popularity.

Experimentation: Fact v. Fiction
Today, doctors continue to seek and discover new uses for oxygen therapy. Clinical testing of hyperbaric oxygen therapy has been used to destroy disease-causing microorganisms, assist cancer patients, tackle chronic fatigue syndrome, and alleviate allergies.

Some studies go so far as to claim that hyperbaric oxygen therapy may benefit patients with HIV, arthritis, sports injuries, multiple sclerosis, tinnitus, cerebral palsy, senility, cirrhosis, Lyme disease, gastrointestinal ulcers, and childhood autism. But while the list of ailments treatable with modern hyperbaric oxygen therapy is long, supporting scientific evidence is spotty. The lack of proper protocol in some clinical studies makes it difficult to determine its true value for some suggested uses.

Still, in the U.S., hyperbaric oxygen therapy is accepted by Medicare, Medicaid, and some commercial health insurers as a reimbursable treatment for many approved conditions. Doctors often prescribe it to address the effects of such diverse conditions as stroke, slow-healing skin lesions, and migraines, as well as a host of diabetes-related conditions such as diabetic retinopathy,

CHAPTER 9: HYPERBARIC OXYGEN THERAPY

diabetic nephropathy, and recurrent foot wounds.

As a general rule, if your health insurer will reimburse a doctor for the treatment, it no longer qualifies as "experimental." If your health insurer won't cover treatment, it's probably because the treatment lacks clinical testing for that use.

Potential Risks

While hyperbaric oxygen therapy is a common procedure, as with any medical treatment there are certain risks. Some individuals with histories of claustrophobia may have trouble with the confinement of the tank, while the change in pressure may negatively impact patients suffering from sinus, eardrum, and lung-related issues (similar to riding in an airplane or deep-sea diving with a stuffy nose). More serious complications can include myopia (short-sightedness that can last for weeks, and on rare occasions, longer), sinus damage, ruptured middle ear, and lung damage.

Rarely, a complication called oxygen toxicity may result in pulmonary effects or seizures. "High risk" patients, such as those listed above as well as patients suffering from acute carbon monoxide and cyanide poisoning, acute crush injury (trauma resulting from falls, car crashes and gunshot wounds, for example), smoke inhalation or closed head injuries (e.g., concussions) are often given "air breaks," during which ordinary air is substituted for pure oxygen at intervals throughout treatment.

Patients with severe congestive heart failure may have negative reactions to hyperbaric oxygen therapy, while those with certain types of lung disease (including those already dependent upon a ventilator for regular breathing) may be at higher risk for a collapsed lung. In most instances when these patients receive hyperbaric oxygen therapy, technicians utilize cardiac monitoring devices throughout treatment.

Pregnant women should be treated with hyperbaric oxygen therapy only in serious situations where there are no other options, such as acute carbon monoxide poisoning. Furthermore, due to the use of pure oxygen, hyperbaric oxygen chambers can also be a fire hazard if not properly maintained and monitored.

What's Next?
As hyperbaric oxygen therapy has grown in popularity, so has the technology that makes it possible. Today's medical equipment is much more efficient at regulating oxygen in the bloodstream than treatments in the 1930s. These advances should continue to make such procedures available for additional diagnoses and on a wider basis, and after further testing it's likely that health insurers will expand the number of uses this treatment covers.

CHAPTER 10: THE LANGUAGE OF BLOOD

"Who has fully realized that history is not contained in thick books but lives in our very own blood?"
— Carl Jung, Swiss psychiatrist and founder of analytical psychology

HOSPITAL STAY

10

THE LANGUAGE OF BLOOD

TOPICS: Blood, Diagnosis, Donor, Draw, Hemoglobin, Hypodermic Needle, Phlebotomist, Platelet, Syringe, Vein

Blood. Few words have the same versatility and dramatic effect. For some, the word calls to mind vampires like Dracula. Others picture American Red Cross workers tending to the wounded. Those in the medical field think of something altogether different. For them, blood means information.

Blood is the human body's life source, providing a continual supply of nutrients and oxygen. Unlike other critical elements that give life to the human body, such as air and water, the thought or sight of blood often makes people uneasy. One person who isn't bothered by this substance is your hospital phlebotomist (pronounced "fli-BOT-o-mist") whose job it is to draw blood.

Almost every morning, a hospital patient can expect to be greeted by breakfast and a visit from this odd-sounding specialist—the person whose job it is to draw blood. In hospitals around

Chapter 10: The Language of Blood

the world, the most common diagnostic tool of health care professionals is the blood test. Blood tests can evaluate your overall state of health and help pinpoint the medical basis for hospital admission. A single sample can help your doctor assess the functionality of major organs, look for potential diseases and conditions, and follow the progress of any treatments that have begun. Blood tests play an important role in helping your doctor determine your length of stay in the hospital. All this, from a tablespoon of liquid.

How a Blood Draw Works

Blood draws are normally short procedures, taking only a few minutes either in a doctor's office, hospital, or laboratory. Depending on the test, you may be requested to fast, or abstain from eating or drinking anything but water, for 8 to 12 hours beforehand. This helps to obtain an accurate sample.

A fingerstick sampling, or a quick prick of the finger to collect a few drops of blood, may be used when the quantity of blood needed is minute, but phlebotomists most often perform venipuncture (think of "vein puncture"), which involves inserting a needle into a vein for direct access.

In the United States, most blood draws are performed with an evacuated tube system consisting of a hypodermic needle, a vacuum tube, and a plastic hub to connect them. (See photo on opposite page.) These tubes contain a vacuum that draws the blood directly into them. Multiple tubes may be attached one after the other to a single needle, rather than having to repeat the procedure for each draw.

In many instances, such as with the elderly or patients whose veins are unreliable, a syringe is used to collect the sample. Unlike the vacuum tubes, syringes require a manual draw, which provides added control and places less pressure on the veins. Once the needle is inserted into the vein, a small amount of blood, called a "flash," appears at the back of the needle, indicating a successful

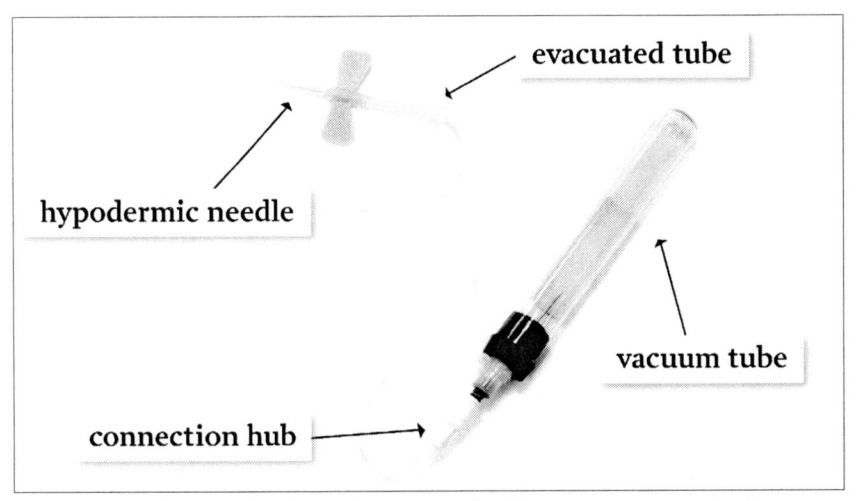

To draw blood, your phlebotomist will probably use an evacuated tube system such as the one shown here.

insertion. At this point the plunger is drawn backwards and the syringe fills with blood.

Depending upon the tests given and the state of a patient's veins, a phlebotomist has many sizes, or gauges, of needle from which to choose. The higher the gauge, the smaller the needle.

What to Expect

Blood is most often drawn from the arm. Your phlebotomist ties a band around your upper arm or asks you to make a fist so your veins swell and become more accessible. Next, your arm is cleaned with an antibacterial agent.

Once the needle is inserted, a small test tube is attached. The tube is filled and then removed, sealing itself in the process so as to prevent leakage or contamination. Should multiple tubes be required, your phlebotomist will continue the process until the desired amount of blood is obtained, at which point the needle is removed from your vein.

A blood draw is usually a painless procedure, though you may feel an initial stick when the needle enters the body. Many people

CHAPTER 10: THE LANGUAGE OF BLOOD

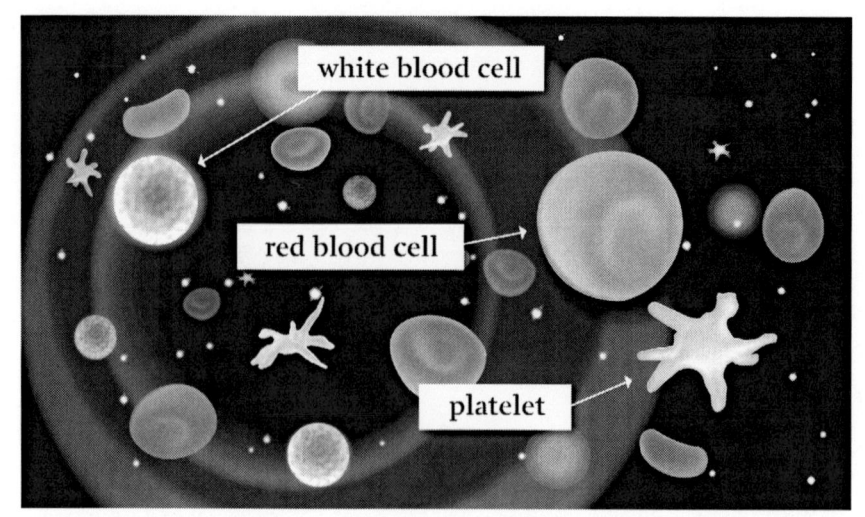

[Illustration: white blood cell, red blood cell, platelet]

Analyzing blood components such as the red and white blood cells and platelets illustrated here can aid in your diagnosis.

don't like to watch, so you might want to look away and distract yourself during the few short minutes it takes to administer the test. Despite what you may think, the risks from a blood draw are minimal. Patients suffering from certain conditions such as belonephobia (fear of needles) and hemophobia (fear of blood) may experience unusual discomfort during a hospital stay. Blood tests are often mandatory, and most people just get used to it.

Once the needle is withdrawn, a bandage or gauze is applied with gentle pressure to stop the bleeding and reduce swelling. At this point, the vials of blood are labeled and sent to a lab for evaluation. Depending on the type and number of tests necessary, your doctor will have results in anywhere from a few minutes to a few weeks.

Types of Blood Tests

Though it's a simple procedure, the blood test is a versatile diagnostic tool and is invaluable to physicians. The information gained from studying your blood allows your doctor to monitor the functioning of your organs and diagnose a variety of potentially fatal diseases such as diabetes, HIV/AIDS, cancer, and anemia. Blood tests are also instrumental in ascertaining your risk for heart disease and determining how well or poorly any treatments and medications are working.

Complete Blood Count (CBC)
The CBC is the most commonly ordered blood test and uses specialized machines to analyze the many components of your blood. A CBC calculates the number of red and white blood cells, as well as platelet, hematocrit, and hemoglobin levels within the blood. This enables your doctor to detect diseases and disorders such as blood cancers, infection, anemia, and clotting issues.

White Blood Count (WBC)
As its name suggests, a white blood cell count, also known as a leukocyte count, detects the number of white blood cells in a volume of blood so that your doctor can compare it to the normal range of between 4,300 and 10,800 cells per cubic millimeter. White blood cells play an important role in the functioning of the immune system and are integral in helping the body ward off infection. A WBC differential also provides "key" information about the blood's granulocytes, lymphocytes, monocytes, eosinophils, and basophils. High WBC counts often indicate infection or malignancy. A low WBC count may signify immune related issues, bone marrow problems, or negative effects from certain treatments such as chemotherapy.

Red Blood Count (RBC)
The red blood cell count, also called the erythrocyte count, calculates the number of red cells in a volume of blood so that your doctor can compare it to the normal range of between 4.2 to 5.9 million cells per cubic millimeter. Smaller than white blood cells

Chapter 10: The Language of Blood

but larger than platelets, red blood cells are the most common cell type in the human body. Their role is to transport oxygen to all the other cells so they can remain healthy and productive. Low RBC counts often indicate anemia and may require further investigation, while high RBC counts may suggest bone marrow disease or low blood oxygen levels.

Hematocrit (Hct) Level

Hematocrit levels provide the ratio of the volume of red blood cells to the overall value of blood. Normal cell ranges differ between males and females, 45 to 52 percent for men and 37 to 48 percent for women. Low hematocrit levels suggest anemia, but may also be caused by conditions such as traumatic injury, bone marrow issues, nutritional deficiencies, and kidney failure. Elevated levels may indicate lung disease or simple dehydration.

Hemoglobin Level

This test measures levels of hemoglobin, protein molecules that give red blood cells their color and transport oxygen throughout the body. A normal hemoglobin range is between 13 to 18 grams per deciliter for males and between 12 to 16 grams per deciliter for females. Low levels of hemoglobin indicate anemia, possibly caused by bone marrow problems, nutritional deficiencies, kidney failure, or the aftereffects of certain treatments. High levels often suggest lung disorders such as emphysema.

Platelet Count

Platelet counts calculate the number of platelets, or fragments of cytoplasm found within cells in the bone marrow, that are vital to carrying out the clotting function. When you're cut, platelets bind together to seal openings in an effort to prevent blood loss. Low platelet counts are often caused by bleeding disorders where blood does not clot significantly, while high counts point to thrombotic disorders, where the blood clots too much, causing clogged vessels and poor circulation.

Mean Corpuscular Volume (MCV)
Mean corpuscular volume measures the average size of your red blood cells. Abnormalities in MCV may signify anemia or thalassemia, an inherited blood disease.

Basic Metabolic Panel (BMP)
The basic metabolic panel is a series of tests that measure existing chemicals within the blood. Focusing on blood glucose (sugar), calcium, electrolyte, and kidney tests, a BMP breaks down the levels of blood chemistry and provides your doctor with detailed diagnostic data pertaining to the health of the heart, liver, kidneys, bones, and muscles. You may be required to fast prior to testing. The following tests make up a typical basic metabolic panel:

- **Blood glucose tests** measure the amount of a specific type of sugar found in blood that the body converts into energy. This test is often used to check for diabetes.
- **Calcium tests** look for abnormalities in the level of calcium within the body to detect possible kidney issues, cancer, bone diseases, thyroid problems, malnutrition, or other disorders.
- **Electrolyte tests** measure the patient's level of electrolytes, minerals that help maintain fluid levels and control the acid/base ratio in the body. They include sodium, potassium, bicarbonate, and chloride. Abnormalities in electrolyte readings may point to dehydration, kidney disease, liver malfunction, high blood pressure, or heart failure.
- **Kidney tests** look for an excess of waste within the bloodstream. The major role of the kidneys is to filter blood urea nitrogen (BUN) and creatinine from the body, and abnormal levels may indicate kidney disease or failure.

CHAPTER 10: THE LANGUAGE OF BLOOD

Blood Enzyme Tests

Blood enzyme tests can be used to detect enzyme levels in the body, and may focus on several different areas such as the heart or liver. Enzymes are chemicals that help to control reactions in the body. Doctors use these tests, often referred to as cardiac enzyme tests or liver enzyme tests, to determine how well certain organs are functioning, as well as to check for possible tissue damage. For example, cardiac enzyme tests are performed to assess the damage done to the muscles of the heart after a heart attack, while liver enzyme tests investigate the extent of liver diseases or disorders.

When muscle or heart cells are injured, creatine kinase (CK) leaks into the blood. Doctors can determine the extent of damage inflicted by the amount of CK found. Troponin is a muscle protein that helps muscles contract. CK and troponin tests are used when a patient is experiencing chest pains or other heart attack related symptoms.

Lipoprotein Panel

A lipoprotein panel is used to evaluate your risk for coronary heart disease. It focuses on the materials in your blood that carry cholesterol, a fatty substance found in the outer lining of cells that plays a large role in vascular health. Fasting is required for 8 to 12 hours before drawing blood for this test.

A typical lipoprotein panel addresses the total cholesterol count, including HDL, LDL, and triglyceride counts.

> **HDL (high-density lipoprotein):** Made of high amounts of protein and relatively small amounts of cholesterol, HDL carries cholesterol through the blood. You may have heard HDL referred to as "good cholesterol," because high levels of HDL lower your risk of coronary heart disease and atherosclerosis by decreasing blockages within the arteries.

LDL (low-density lipoprotein): Made of moderate amounts of protein and high amounts of cholesterol, LDL carries cholesterol through the blood. Because high levels of LDL increase your risk for coronary heart disease due to cholesterol buildup, you may have heard LDL referred to as "bad cholesterol."

Triglyceride count: These naturally occurring glycerides provide energy and form much of the fat stored by the body. They are made up of three individual fatty acids bound together in a single large molecule.

Listen to Your Blood

If your doctor suspects a blockage in your arteries or veins due to atherosclerosis (thickening of the artery walls) or thrombosis (blood clot), a Doppler arterial or venous study may be ordered. This type of test is unique among blood tests because it doesn't require a needle. It's an ultrasound diagnostic test that uses a handheld instrument the doctor can glide along your skin, above the area of the suspected problem. Sound waves are detected and then recorded as they bounce back from the arterial flow, providing your doctor with images of your blood flow to determine the direction and measure the speed at which your blood is traveling.

As a diagnostic tool, the blood test is a powerful ally. Using a single vial of blood, your bodily functions—from organ vitality to muscle and tissue health to vascular strength—can be mapped out and evaluated, enabling your doctor to draw an overall picture of your state of wellness. Normal ranges can differ depending on age, race, and gender, so your doctor will take these factors into consideration when determining whether your test results are normal. Armed with this information, doctors are better able to identify areas of potential concern and evaluate the progress of medication and treatment.

CHAPTER 10: THE LANGUAGE OF BLOOD

If you receive news of an abnormal test, there is no immediate cause for alarm. In addition to diseases and medical disorders, there are many factors that may contribute to abnormal readings. Diet, exercise, alcohol consumption, smoking, menstrual cycle, and recent medications can all have an effect on initial blood test readings. In discussing test results with your doctor, you can gain a better understanding of your body's inner workings and its current state of health.

In this way, the blood test proves its true potential. By revealing critical information found within your biochemical makeup, blood tests assist your doctor and may be able to help discover potential problems early, at a time when lifestyle changes or medical treatment can be more effective.

No matter the reason for your visit, if you find yourself in a hospital, chances are that someone will draw your blood.

WHAT'S YOUR TYPE?

Whenever a transfusion is necessary, doctors must first determine blood compatibility. The most common blood groupings are the "ABO" and "Rh" systems. The ABO system classifies human blood into four major types based on the presence or absence of two antigens on red blood cells ("A" and "B"). Individuals can have one of four blood types: A, B, AB, or O (O is the absence of both antigens). The Rh system classifies blood according to the presence or absence of the Rh antigen, often called the Rh factor. Individuals either have an Rh factor on the surface of their blood cells (indicated by a plus (+) sign), or they do not (indicated by a minus (-) sign). The following chart indicates red blood cell compatibility between two individuals according to both the ABO and Rh systems:

				Donor				
Recipient	O-	O+	A-	A+	B-	B+	AB-	AB+
O-	√							
O+	√	√						
A-	√		√					
A+	√	√	√	√				
B-	√				√			
B+	√	√			√	√		
AB-	√		√		√		√	
AB+	√	√	√	√	√	√	√	√

People with type O- blood are commonly referred to as "universal donors" because any individual can receive this blood (with a few rare exceptions). People with AB+ blood are commonly referred to as "universal recipients" because an individual with this blood type can receive any of the four blood types with either Rh factor (again, with a few rare exceptions).

Chapter 11: The Hospital Menu

"My mother's menu consisted of two choices:
Take it or leave it."
—Buddy Hackett, American comedian and actor

11

THE HOSPITAL MENU

TOPICS: Centralized and De-Centralized Kitchen, Dietitian, Hot Plate, Jell-O, Microwave Oven, Nutrition

Hospital menus across the globe can be as diverse as the languages spoken. In Mexico, your hospital stay may include steak or chicken, beans and rice, and corn or flour tortillas. In China, hospital cuisine may include rice, eggs, meatballs, vegetables, and soup. In the U.S., you can usually count on chicken, potatoes, and a vegetable. But no matter where you are in the world, hospital patients have one thing in common—no one comes for the food. Still, studies show that what we eat plays a critical role in our health and the healing process, and hospitals have responded.

Today's hospital meals have come a long way from Jell-O cubes, thanks to advances in technology and nutrition. Surprisingly, architecture also bears a certain responsibility for what you find on a hospital menu. Together, these factors have helped to shape hospital meal preparation as it has adapted to changing environments.

Chapter 11: The Hospital Menu

As you sit in your hospital bed, consider the challenges involved in preparing dinner for any number of patients with different illnesses and nutritional needs. Recognizing what it takes to keep patients nourished and satisfied may make the cuisine more palatable, which will certainly improve your hospital experience.

A Historical Perspective

The need to feed the sick has played a role since hospitals first came into being. *Hospital* comes from the Latin word for "host," and early hospitals dealt as much with caring for the poor and needy as they did ministering to the sick. It was not until the eighteenth century that the concept of the strictly medical-based hospital as we know it today came into being in the U.S.

For most of the twentieth century, hospital menus—like the food served in public schools—were high in fats (often from red meat), refined carbohydrates (white bread, white rice, corn syrup, table sugar), processed foods (American cheese products, store-bought cookies and crackers, etc.), and low in fiber (whole wheat, brown rice, bananas, broccoli, etc.). With our current knowledge of nutrition, it's surprising to note that while hospitals were boasting of such medical advances as antibiotics, a vaccine for polio, and non-invasive surgery, few were mindful of the menu and the ways in which it related to patient care. As hospitals came to grips with the importance of nutrition, the evolution of hospital design and building technology began to influence how the hospital menu could be composed and executed.

The preparation, cooking, and serving of food in a hospital began as a decentralized process. In the early 1900s, when most hospitals consisted of low, sprawling pavilions, food was delivered in large cans to be reheated and spooned out in the wards as needed. The introduction of hot plates made it easier to heat these simple meals outside of the kitchen, but hospital staff still struggled to keep the food warm as it traveled to patients.

By 1930, the age of multi-story hospitals had arrived, and hospitals went back to relying upon centralized kitchens, serving individual trays and food packs for private patients. As the increase in population continued to influence hospital growth, most medical centers expanded again—this time, outward rather than up. These new, sprawling hospital campuses were not well served by centralized kitchens, and within a decade hospitals reverted to the decentralized model.

It was not until the invention of the microwave oven in 1945 that the hospital menu was finally able to overcome its architectural challenges. The microwave allowed kitchens to institute "cook-chill" service, a process wherein food is cooked, frozen, and then reheated as needed. This enabled kitchens to increase variety while maintaining quality. Patients suddenly found themselves being treated as individuals whose daily meals drew nearly as much emphasis as their treatments.

Technology and Nutrition Come Together
With the support of today's modern kitchens, doctors can help patients not just by prescribing medication, but also by customizing hospital meals to complement a patient's care. Most U.S. hospitals now employ registered dietitians to ensure that patients eat a variety of healthy foods during their stay and learn about basic nutrition so they can maintain a proper diet at home. To use the title "Registered Dietitian," these health care professionals must meet strict educational and professional prerequisites, regulated by the American Dietetic Association (ADA), and they must also pass a national registration examination.

Dietitians create menus that meet healthy eating guidelines set by the ADA and satisfy regional tastes. They also address the needs of patients with dietary concerns such as diabetics, breast-feeding moms, and those with food-related allergies. Dietitians also provide "medical nutrition therapy," which includes tube

Chapter 11: The Hospital Menu

feedings (enteral nutrition) and intravenous feedings (parenteral nutrition). Some clinical dietitians have dual responsibilities that focus not only on patient care, but food planning and service for the entire hospital as well.

When a patient is admitted, a dietitian will review his or her medical history to determine the most appropriate diet. Screening hospital patients for possible food allergies and keeping a watchful eye out for medically incongruous meals is critical, as this can significantly affect a patient's current diagnosis and course of treatment. Thankfully, dietitians today provide the food-related clinical care throughout most hospital stays.

New Trends for the Hospital Menu

The concept of food as preventive medicine has resulted in some leading hospitals offering primarily organic and chemical-free food, including hormone-free milk, antibiotic-free chicken and beef, and locally grown fruits and vegetables. According to a recent article in Time magazine, Kaiser Permanente, one of the nation's largest nonprofit health systems, sponsors weekly farmer's markets at 29 of their hospitals and offers milk from cows raised with no synthetic hormones at its medical centers. By doing so, Kaiser hopes to educate patients about healthy lifestyles by urging them to consume more locally grown, unmodified fruits and vegetables.

Today, nearly 20 percent of American hospitals outsource their food service in one form or another, and this trend is on the rise. By leaving the creation of nutritious dishes to the culinary experts, hospital personnel are now free to focus more on the patients. If implemented properly, outsourcing food service can increase hospital efficiency, improve service, and offer better food with a greater selection, all resulting in higher patient satisfaction.

Raising the bar even higher, many outsourcing companies now offer room service dining, similar to hotels. Patients simply place their order with the kitchen, and the request is delivered in a timely fashion. The meals are all freshly prepared and the menus are extensive.

As menus and the responsibility for meal preparation have expanded, new computer software has made it easier to track a patient's dietary restrictions and allergies, as well as individual likes, dislikes, and/or religious concerns. Although most patients are healthy enough to order whatever they like, these new systems alert hospital staff if a patient places an order that is inconsistent with his or her diet plan. These alerts also help to educate patients on the importance of maintaining healthy diets outside of the hospital setting.

Food as Medicine

As always, the evolution of medicine continues to take its cue from Hippocrates, who must have thought holistically when he said, "Let food be thy medicine and medicine be thy food."

Nowadays, hospitals no longer tout the benefits of healthy eating while serving meals lacking in both nutrition and taste to their patients. With revamped menus, patients are not only eating food free of excess sugars, starches, and preservatives, but they are eating more of it. Recent surveys show that patient satisfaction increases as a result of better fare, and more patients are eating their entire meal and enhancing their nutritional intake, which in turn can speed healing time. Today's medical menu not only assists healing in the short term, it also seeks to teach patients how to recalibrate their eating habits to promote better health in the future.

Chapter 12: Building a Balanced Hospital

"I recently went to a new doctor and noticed he was located in something called the Professional Building. I felt better right away."

— George Carlin, American comedian, actor, and author

12

BUILDING A BALANCED HOSPITAL

TOPICS: Bacteria, Incandescent and Fluorescent Lamps, Mercury, Seismic Safety, Volatile Organic Compounds (VOCs)

Nobody likes a dirty hospital. It's the fantasy of every patient to be greeted by a smiling nurse in a crisp uniform, then whisked away to a private room of glowing white and gently placed in a bed of gleaming chrome and spotless, comfortable sheets. This hygiene, however, comes at a price.

Hospitals pollute in their effort to help the patient. Chemical toxins (found in everything from cleaning products to radiology equipment), heavy energy consumption (essential for running a hospital 24 hours a day, 7 days a week), and medical waste are inevitable hospital by-products.

But does "good for the patient" have to mean "bad for the planet?" Hospitals across the country are beginning to adjust their practices as they recognize the patient-care and fiscal benefits of ending their polluting ways. But it's not so easy being green when

Chapter 12: Building a Balanced Hospital

lives are at stake. In seeking change without sacrificing quality of care, hospitals face considerable odds in their bid to reduce their environmental footprint.

Unlike most businesses, whose bottom lines focus on profit, the medical industry sets the health of patients as top priority. That gives health providers less flexibility, because any changes to practices must involve no health-related drawbacks. Whether it's maintaining a sterile environment without the use of potentially dangerous chemicals or building safer, more energy-efficient facilities, hospitals need to be cautious with their green efforts.

On top of this, they must do so within the confines of a heavily regulated industry, contending daily with government groups, private sector oversight agencies, and policies that, while well intentioned, may stand in direct opposition to the needs of both the hospital and its patients.

Cleaning Up the Clean-Up

Every day, hospitals must deal with the irony that the products used to maintain a sterile facility are often as dangerous as the germs they eradicate. When it comes to chemicals, less is often more, and there is no question that reducing toxic ingredients in any area will promote a stronger healing environment.

For a hospital in particular, using fewer toxic cleaning products reduces the stress on everyone, from patients to visitors to employees, while improving overall safety and making for a more health efficient facility. But what are the alternatives, and what are the risks?

Traditional cleaning products are toxic and full of VOCs (volatile organic compounds). VOCs, found in an estimated 35 percent of traditional cleaning products on the market, can cause blindness and very serious irritations to the skin as well as internal organs. Long-term exposure to such chemicals has been linked to cancer, damage to the reproductive organs (which can cause infertility), kidney failure, and neurological damage, among other concerns. Even in small doses, the presence of VOCs has been shown to have adverse effects on the healing process.

Despite such dangers, potent cleaning products are needed to maintain a sterile environment and reduce risk of infection. A hospital's green cleaning program must involve a multi-disciplinary team for research, implementation, and oversight.

The result is a fine line, as too much disinfection can cause harm to patients, visitors, and employees, while not enough can lead to hospital-acquired infections, for which the hospital must shoulder the blame.

Phasing Out Mercury

In recent years, most hospitals have made significant progress in eliminating mercury from their facilities. Both versatile and dangerous, for years mercury played an important role in all sorts of clinical devices, from thermometers to thermostats.

While mercury's usefulness is beyond question, it is also a powerful neurotoxin with the potential to affect the brain, spinal cord, and major organs. That's why hospitals throughout the United States are replacing mercury-containing devices, sometimes at significant expense.

To begin, hospitals must identify and eliminate mercury-containing laboratory equipment and replace them with efficient alternatives. While devices like fever thermometers, laboratory thermometers, wall blood pressure units, laboratory tubes, and foley catheters have historically held small amounts of mercury, some of the larger mercury-containing devices include fluorescent light bulbs, thermostats, mechanical/tilt switches, flow meters, flame sensors, gas meters, and boiler gauge controls (some boilers can contain up to 75 pounds of mercury).

Hospital laboratories use other dangerous materials, too. Histology and cytology procedures, where lab technicians work with

Chapter 12: Building a Balanced Hospital

microscopic cells, require the use of staining reagents to distinguish cells. Common staining reagents have few available nontoxic alternatives.

Toxins in Electronic Equipment
Medical practitioners also find themselves surrounded by potentially harmful diagnostic tools. The advanced technology needed to properly scan and monitor patients—computers, televisions, lab analyzers, EKG monitors, and other types of biomedical equipment—often contain hazardous materials.

Whether it is the lead (even the ancient Romans were aware of the metal's ill effects) in a cathode ray tube (CRT) monitor, the toxic plastics in cable wiring, brominated-flame-retardant-infested circuit boards, or the mercury in LCD displays, health care facilities must be vigilant in removing equipment that contains toxins. It is in any hospital's best interest to effectively minimize the prevalence of dangerous substances without affecting the quality of available technology.

Without an appropriate disposal protocol, hospital waste created by updating equipment and resources can threaten the environment and public health. Hospitals must also be mindful that their disposal practices comply with a complex network of federal, state, and local laws specific to health care.

In a medical setting, technical upgrades are not so simple as changing a printer cartridge, and proper disposal must be used to ensure that these improvements don't have a negative environmental impact. It is the hospital's responsibility to make sure its environmental cure is not worse than the disease.

When the Lights Are Always On
Medical emergencies don't follow a set schedule. Between the constant demand for 24-hour ER services and round-the-clock monitoring of admitted patients, hospitals are one of the few

places where the lights are kept on all day, every day.

American hospitals use more than double the energy of office buildings of the same size, and twice the amount of European hospitals. This makes them a significant source of greenhouse gas emissions, which some scientists believe is contributing to climate change (also referred to as "global warming").

One thing is for certain: lights in hospitals are a necessity. So, what can be done to reduce the burden? For starters, the transition from incandescent lamps to fluorescent lamps reduces energy consumption and ensures a longer usage time, benefiting both the environment and the hospital's bottom line. But even such a seemingly simple act of progress comes with strings attached, for if not disposed of properly, fluorescent lamps may release mercury (yes, mercury again) into the air, water, and soil.

Technology, too, places a strain on a hospital's energy bill, and with advances in medical equipment and procedures come the added costs of maintaining them. In its quest for greater sustainability, the modern hospital must further contend with the conundrum of providing its patients with state-of-the-art health care without raising its energy use unduly.

Identifying the problem is straightforward: Hospitals use too much energy. Yet the entire industry must consider the effects on the safety of its patients before taking any action toward reducing overall consumption.

Building a Greener Hospital

Hospitals across the country are beginning to realize that being eco-friendly makes good sense, both financially and in terms of patient care. Some hospitals are eliminating disposable water bottles, while others are purchasing hybrid vehicles and switching to energy-efficient lighting. But for many, the answer lies in the construction of new facilities that are both safer and more energy efficient in the long run.

CHAPTER 12: BUILDING A BALANCED HOSPITAL

The state of California provides a few excellent examples of the future of modern medical buildings. After an earthquake measuring 6.7 on the Richter magnitude scale caused structures in and around the Northridge neighborhood in Los Angeles to collapse, legislators asked all hospitals in the state to meet seismic safety standards. As a result, California hospitals will spend about $120 billion in order to meet these standards, and several institutions have seized on this opportunity to incorporate a green vision into their retrofitting projects.

For example, Mills-Peninsula Health Services in Burlingame, California, is launching a new facility just two miles from the San Andreas fault line, spending more than $500 million to be perhaps the most seismically safe hospital in the country. The hospital will also come complete with cool roofs (which can save on cooling and heating), low-VOC materials and finishes, furniture and other fixtures made from recycled materials, high-powered solar energy to make things warm and cold, and a ventilation system relying upon fresh air.

Brand new facilities that replace older hospitals are sprouting up all over the state, though sadly very few focus on environmental upgrades. One exception to this pattern is Kaiser Permanente, which has an environmental committee that influences its construction decisions. Kaiser plans to spend close to $24 billion over the next several years, and 30 million square feet of this new construction will come from what the company calls "ecologically sustainable materials." Another bonus is that because Kaiser has dozens of medical centers nationwide and is buying sustainable materials in such large quantities, the company has helped to drive down costs.

In the future, green hospital construction should have a beneficial impact on both the environment and the patient. Still, good intentions are not enough, and caution must be used when implementing such changes. Increasing natural airflow has been shown to spread airborne diseases, while reducing the use of water too much can increase hospital infections. Still at the forefront, Kaiser is testing a displacement ventilation system that, the company says, "introduces air at floor level and uses the natural buoyancy of warm air to remove particulates," which may reduce energy expenses by up to 40 percent. Kaiser has confirmed that this system will not be used until it is found to be both safe and effective.

Overall, it is in a hospital's best interest to clean up its environmental footprint, no matter how difficult competing influences may make the process. Effectively pursuing the transition to a greener hospital is possible, despite the initial cost, risk, and inconvenience, and going green may add to greater efficiency and savings down the line.

CHAPTER 13: ABOUT ASPIRIN

"Poisons and medicine are oftentimes the same substance given with different intents."
— Dr. Peter Mere Latham, British physician and educator

13

ABOUT ASPIRIN

TOPICS: Acetylsalicylic Acid, Blood Clotting, Food and Drug Administration, NSAIDs, Reye's Syndrome, Willow Tree Bark

If you find yourself in the hospital for heart disease or cerebrovascular disease, it's likely that your doctor will prescribe acetylsalicylic acid, more commonly known as aspirin.

This over-the-counter drug (meaning you don't need a doctor's prescription to buy it) has been available in one form or another since Greek physicians first boiled the bark of the willow tree to combat fever and aches nearly 2,500 years ago. Since then, aspirin has undergone a number of improvements to become one of the world's most widely used drugs. Not only does it relieve pain, reduce inflammation, prevent blood clots, and lower fever, but this most unusual of compounds is also said to cure acne, tackle dandruff, improve your garden, and remove stains.

Recent studies indicate that more than 50 million Americans (or 36 percent of the adult population) consume 10 to 20 billion

CHAPTER 13: ABOUT ASPIRIN

low-dose aspirin tablets, capsules, and caplets each year for preventative cardiac care alone. With such a diverse set of healing properties, aspirin is not only a miracle drug, it is also big business. The worldwide sale of 50 billion tablets annually translates into about $600 million. Be it medical or financial, this is one powerful pill.

The Evolution of Aspirin

Aspirin as we know it is the result of a succession of steps designed to maximize the healing properties of salicin, a naturally occurring compound found in the leaves and bark of the willow tree. Today, each tablet consists of a mixture of four simple ingredients: the active ingredient (about 325 milligrams of acetylsalicylic acid), corn starch, water, and a lubricant—such as vegetable oil.

From bark to bottle, however, was not as direct a path as one might think. First sold in powder form, the aspirin of today owes much to the skills of those early chemists and researchers who saw the potential of salicin (a naturally occurring compound similar to acetylsalicylic acid) and continued to make improvements over the years.

Experiments with what was later to be known as aspirin began in Europe in the early nineteenth century. Initially, when the willow leaves and bark were boiled, the process produced purified salicylic acid, an analgesic, but it was bitter and irritated the stomach when taken orally.

To counteract these side effects, scientists aimed to neutralize the substance and make it taste better. In 1893, Felix Hoffman and Arthur Eichengrun, chemists for Friedrich Bayer and Company, were able to remove all negative effects of earlier trials by buffering the compound and making it easier on the stomach. Though the taste was still bitter, all medicinal benefits remained intact. Aspirin was born.

WHO SHOULDN'T TAKE ASPIRIN?*
* Unless directed by a doctor
1. Children and teens under 17
2. Women who are pregnant or nursing
3. Patients of any age with a viral infection
4. Patients with gastrointestinal bleeding
5. Patients with allergies to ibuprofen or naproxen
6. Patients with tumors
7. Patients who have headaches of unknown cause

Aspirin Today
While many other pain relievers have over time flooded the market under brand names like Tylenol, Advil, and Aleve, aspirin is still the only pain reliever/fever reducer with demonstrated beneficial effects for heart attack and stroke patients. As a result, home consumption is a regular part of everyday life for many Americans. Doctors prescribe aspirin for headaches, menstrual pain, minor arthritic pain, muscular pain, fever, and toothaches.

But unlike other pain-relieving medications, studies have conclusively shown that low doses of aspirin inhibit the formation of blood clots, leading the medical world to recommend a daily tablet for those at risk for heart attack or stroke. Further experiments are underway to determine whether this multifaceted pill is effective in deterring certain cancers, namely those of the colon and lung, as well as preventing liver damage.

Caution: Side Effects of Aspirin
Even over-the-counter drugs come with a host of warnings. While aspirin is by most accounts a relatively safe, non-addictive drug, like all medications it should be used wisely and only as directed. Because aspirin is available without a prescription, it is considered by many to be harmless, often resulting in dosage-doubling in an attempt to enhance its strength or speed its effect.

CHAPTER 13: ABOUT ASPIRIN

Aspirin can also have potentially harmful side effects, even in low doses. Though much improved over the years, aspirin is still an irritant to the stomach lining. Both doctors and manufacturers recommend avoiding the drug if you are at risk for peptic ulcers, gastritis, alcoholism, or diabetes. In fact, nearly one-third of all gastrointestinal bleeding related hospitalizations and deaths are the result of using aspirin, ibuprofen, or other NSAIDs (nonsteroidal anti-inflammatory drugs). Furthermore, aspirin's ability to slow blood clotting, so highly regarded for heart attack and stroke patients, may actually work against the patient with a gastrointestinal bleed.

Aspirin also restricts the kidneys' ability to excrete uric acid, which may lead to complications for patients suffering from kidney disease, hyperuricemia, or gout.

Children and teenagers are no longer advised to use aspirin for flulike symptoms due to a much publicized link to Reye's syndrome, a condition causing fatty liver and encephalopathy (brain disease). In fact, people of any age are at risk of developing this condition if they take aspirin during a viral infection. Women who are pregnant or nursing are also warned to avoid aspirin, as aspirin can be passed to the fetus or baby. Aspirin should not be taken by people who have a known allergy to ibuprofen or naproxen, or who exhibit signs of salicylate intolerance such as severe headaches accompanied by hives and swelling.

Interestingly, another side-effect of aspirin is the very reason some physicians prescribe it—and it can be very hazardous to your health. Aspirin hinders platelet formation, thereby slowing or inhibiting the body's ability to form clots. For this reason, aspirin is believed to be beneficial in preventing heart attacks and strokes. However, hindering clot formation can be extremely dangerous when a patient is bleeding internally from a stroke (often evidenced by a simple headache). In these cases, aspirin use can rapidly worsen cerebral bleeding and, just as with gastrointestinal

IS IT ASPIRIN?

Many people call all over-the-counter pain medications "aspirin," even though several don't actually contain the compound itself. It's important to consult your doctor when taking any drug, and you should always know the ingredients. This applies to both prescription and OTC medications.

Common OTC Pain Relievers and Their Active Ingredients

Brand Name*	Aspirin	Acetaminophen	Ibuprofen	Naproxen Sodium
Bayer	✓			
Excedrin	✓	✓		
Ecotrin	✓			
Tylenol		✓		
Anacin-3		✓		
Aspirin-Free Anacin		✓		
NyQuil		✓		
DayQuil		✓		
Feverall		✓		
Advil			✓	
Motrin IB			✓	
Nuprin			✓	
Nurofen			✓	
Aleve				✓

* Due to potential risks associated with ingredients found in many brand name medications, you should always take the appropriate drug consistent with your symptoms and medical history. Take any medication only as directed.

CHAPTER 13: ABOUT ASPIRIN

bleeding, make the condition much worse as the body's mechanisms for clotting have been interrupted. It's important to remember that even a "benign" medication such as aspirin can have harmful effects when used for the wrong reasons. To be safe, you should always check with your doctor before taking any medication.

People who currently take prescription medications need to consult with their doctor because aspirin has been proven to interact negatively with certain drugs such as anti-diabetic, immunosuppressant, and non-steroidal anti-inflammatory medications.

How Aspirin Is Made
Today the production of aspirin is a thriving, highly technological business, complete with research and development labs, automated manufacturing processes, and public relations departments. The steps that go into making aspirin, however, are not so complex.

To produce the hard aspirin tablets found in most medicine cabinets, manufacturers begin with the active ingredient, acetylsalicylic acid, and add a combination of corn starch and water as both a binding agent and filler, followed by a lubricant. Binding agents keep the tablets in a solid shape rather than powder form, fillers extend the size and shape of the tablets to make them more manageable, and lubricants such as vegetable oil, stearic acid, talc, or aluminum stearate ensure that the mixture does not stick to the machinery during production. To enable tablets to taste better and dissolve more quickly, chewable aspirin contains different lubricants such as mannitol, lactose, sorbitol, sucrose, and inositol, as well as flavor agents such as saccharin.

As with all drugs, the Food and Drug Administration requires that aspirin manufacturers maintain excellent standards when it comes to quality control. All machinery within each plant is thoroughly sterilized before production begins to ensure that the product is not contaminated or diluted in any way. In addition, operators perform periodic tests designed to calibrate and maintain an accurate and even dosage throughout the production process. Tablet thickness and weight are also controlled and regulated. Once produced, tablets are subject to quality tests to determine their ability to withstand the rigors of packaging and shipping and to dissolve at the required rate.

So much goes into the creation of this little pill we all take for granted. With such a colorful history, the aspirin tablet stands in many ways as the perfect representative of the marriage between medicine and manufacturing. Though aspirin may have ceded its popularity as a pain reliever to acetaminophen and ibuprofen, this versatile compound once made by boiling wood continues to adapt to the times and make itself useful.

CHAPTER 13: ABOUT ASPIRIN

ASPIRIN AND THE FDA

Just because you can buy over-the-counter medications without a doctor's prescription doesn't mean they are harmless. It's important to understand the risks and benefits of any drug. The Food and Drug Administration (FDA) oversees the advertising and packaging of medications, including all drug labels. This agency also monitors and approves new drugs for public consumption. Taken together, this oversight helps patients make informed decisions before beginning treatment.

A division of the U.S. Department of Health and Human Services, the FDA regulates OTC drugs based on their active pharmaceutical ingredients. Drug makers include additional ingredients to create unique formulas in an effort to improve taste, make pills easier to swallow, and make them dissolve at the desired rate. Pick up a bottle of aspirin and you'll see "acetylsalicylic acid" listed as the "active ingredient"—meaning the ingredient that makes you feel better. The FDA has approved nearly 800 ingredients which, when combined in different ways, constitute more than 100,000 OTC medications available in the United States today. Many OTC drugs such as Benadryl, Claritin, and Zantac once required a doctor's prescription, but have since been reclassified due to continued safety assessments over time.

Though the Federal Trade Commission (FTC) is technically in charge of monitoring the advertising and promotion of OTCs, recent years have found that the FDA's opinion holds great influence in promoting change when drug manufacturers stray into questionable areas involving public safety. In April 2009, the FDA played an integral part in forcing the makers of OTC pain relievers and fever reducers to revise their labeling on both outside packaging and bottles to include warnings about potential product safety risks. These warnings addressed non-steroidal anti-inflammatory drugs (NSAIDs) such as aspirin, ibuprofen, naproxen, and ketoprofen, which studies have shown can cause stomach bleeding in people who take the medication for longer than directed, in greater doses than directed, or mix the products with alcohol or another NSAID. Acetaminophen, too, has been linked to severe liver damage in patients who take multiple products containing the drug, mix the drug with alcohol, or exceed the recommended daily dosage.

The FDA has also led an ongoing struggle to convince the makers of aspirin not to market their products to children, while alerting the general public to the dangers of administering aspirin to those less than 17 years old. When given to children under 17 who have viral infections such as chicken pox or the flu, aspirin increases the risk of developing Reye's syndrome, a disease which can affect every organ and may result in long-term neurological issues, coma, or even death. The FDA also recommends that children under the age of two should not be given OTC cough and/or cold medicines, which are linked to instances of rapid heart rate, convulsions, unconsciousness, and sometimes death. While the FDA does not have the legal authority to force a pharmaceutical company to discontinue its marketing practices, the agency's stern recommendations in the interest of public health can be very convincing.

When giving any OTC medication to a child, the FDA stresses the need to weigh the benefits of treating the child's symptoms against potential risks. Since children process drugs at a different rate than adults, they may have different adverse reactions. For this reason, the FDA emphasizes the need to thoroughly read medication labels and familiarize yourself with all active ingredients and their potential side effects before administering any medication to a child.

Chapter 14: Understanding Your Hospital Bill

"When at last we are sure / You've been properly pilled
Then a few paper forms / Must be properly filled
So that you and your heirs / May be properly billed."

— Dr. Seuss, aka Theodor Seuss Geisel, American author

14

UNDERSTANDING YOUR HOSPITAL BILL

TOPICS: Explanation of Benefits (EOB), HMO, Insurance, Medicaid, Medicare, Networks, PPO, Premium, Uninsured

Let's face it, hospital stays are full of unfamiliar terms and confusing procedures. From wearing a hospital gown and settling into a small room that you may share with a stranger to understanding your diagnosis and treatment, many patients feel overwhelmed. To make matters worse, your confusion may continue even after you have been discharged and sent home. Weeks pass, life returns to normal, and then one day there it sits, waiting in the mailbox—that jumble of lines, abbreviations, and numbers that can only mean one thing: your hospital bill.

The typical bill can be difficult to understand, even for those in the health care industry. This is because the act of consolidating fees for the various services you receive while in the hospital can be challenging. Any hospital stay will include several services from the hospital's many departments. Most likely you will

CHAPTER 14: UNDERSTANDING YOUR HOSPITAL BILL

receive multiple bills from your hospital stay: one from the hospital, one from the emergency room, and one from the lab. To add to the confusion, your bill will likely contain an abundance of codes and abbreviations, which vary from hospital to hospital and insurer to insurer. In the end, your statement may look more like a foreign-language puzzle than a bill.

How Your Bill Is Calculated

The first step in deciphering your hospital bill is to understand how the health care industry tallies costs. Every time you visit a medical facility, whether it's an ER, clinic, full-service hospital, or doctor's office, a bill is drawn up. Your bill for each visit will include fees for each procedure performed, medication given, and all services rendered during that stay. Each line will likely include information such as:

- The date the good (e.g., medication, food, etc.) or service (e.g., doctor's exam, nurse consult, etc.) was rendered
- The department from which it came
- A brief description of the good or service
- The quantity of the good or service
- The amount billed
- And, in some cases, the insurance "write-off," which is the amount that your insurer (if you have one) has negotiated with your hospital and/or doctor to waive

A typical line from a hospital bill may look as follows:

Date	Rev	Dept	Charge	Description	Qty	Amount
08/24/10	271	4050	1538	Cotton Balls, Pkg	1	$XX.XX

Your Explanation of Benefits (EOB)

If you've provided the hospital with your insurance information, it should automatically bill your insurer on your behalf. The amount for which you are responsible and the amount your insurer pays is typically determined by your policy contract. Once your insurer has reviewed and processed your "claim" (another word for the detailed bill the hospital sends the insurer), your insurance company usually mails you an Explanation of Benefits (EOB). This is what many patients erroneously refer to as their "bill." It is not. Rather, this important document breaks down your hospital bill by explaining which fees your insurance policy will pay. If you plan to deduct medical expenses on your taxes, or if you participate in your employer's medical flexible spending account (FSA) plan, an optional program that enables employees to pay for medical costs with tax-free dollars, you may want to save your EOB for your records.

Your EOB is a summary of:

- Costs that your insurance company believes are (or are not) covered by your policy
- How much your insurance company intends to pay
- How much they intend *not* to pay
- Why certain items are not covered
- How much of the bill is your responsibility

Your EOB explains all of this using procedural codes and fees that the hospital has billed, as well as a line-by-line appraisal of what the insurance company has covered and what remains to be addressed out of pocket by the patient. You'll usually receive your bill from the hospital before you receive your EOB from your insurer. Many patients wait to receive an EOB before they pay their bill because the EOB can be a better indicator of what is actually owed to the hospital.

Chapter 14: Understanding Your Hospital Bill

How Costs Are Determined

Now that you know how to read your bill and EOB, you may wonder how hospitals and insurers determine these amounts. How do they assign monetary values to such varied goods and services as the number of aspirins taken, bed sheets used, or number of minutes examined by a physician in an operating room or spent during a scan?

Most hospitals work with outside companies to determine the average retail price of each good and service provided. They then set their own prices based on these values, as well as prices around the community.

In the case of your insurer, the amount your policy covers depends on the policy you have. Typically, the higher the cost of your monthly "premium" (the amount you and/or your employer pay each month to your insurer), the more your insurance plan is likely to cover, though this varies based on many factors, including your age, general health, pre-existing conditions, and whether you hold an individual or group policy. Insurers have access to vast amounts of cost data, which they study and use as a basis to determine what they think is a fair market value for the goods and services you receive at the hospital. Keep in mind that many insurance companies are for-profit, so they are seldom willing to pay retail hospital charges.

Insurers may also negotiate with doctors' groups and hospitals to pay less for some things in exchange for encouraging patients to go to them. This is where you may have heard the phrases "in-network providers" (health care professionals who have signed

SMART TIP: Emergencies aside, some insurance plans will pay only if you go to an in-network hospital, while others will pay a partial amount if you go to an out-of-network hospital. Whenever possible, contact your insurance company to determine what your policy covers.

contracts agreeing on prices) and "out-of-network providers" (professionals who have not contracted with your insurer). This downward pressure on prices can leave the uninsured out in the cold, as they are charged, and may be obligated to pay, all costs incurred at the hospital without a discount. If you find yourself in such a situation, it's a good idea to speak with the hospital billing department or your doctor, as they may be able to provide a more affordable solution for your care.

A Typical Hospital Bill

To fully understand the process that goes into tallying up a hospital bill, it is first necessary to understand a patient's condition and subsequent treatment as it is divided throughout the hospital. Most major ailments incorporate services from several different departments. A typical hospital bill may not group or order its items by department, though it may tell you from which department each charge originated.

Take a case of pneumonia, for example. If the disease progresses enough to make hospitalization necessary, it can lead to a very expensive bill. The resources needed to effectively combat the infection draw from nearly all factions of a hospital, including the pharmacy, laboratory, radiology, and respiratory departments. To address the different symptoms, a typical pneumonia stay at a hospital often draws a variety of doctors into the equation, such as an internal medicine doctor, an infectious disease specialist, and a pulmonologist.

As a result, the combined hospital charges for treating pneumonia may reflect not just the hospital bill, which may or may not include the emergency department, but also a bill from the laboratory, the radiologist, and the multidisciplinary team who worked to rid the patient's body of its bronchial havoc.

Chapter 14: Understanding Your Hospital Bill

How It All Adds Up

Here's how a bill might be tallied for pneumonia: First, to diagnose your condition, your doctor will likely request an X-ray or CT scan to confirm the presence of pneumonia in your lungs. If your X-ray appears normal but your doctor still suspects bacterial pneumonia, you may be asked to take a blood test or sputum culture (a substance that is expelled from the respiratory tract, such as mucus or phlegm, mixed with saliva).

If an X-ray or CT scan is involved, expect to visit the radiology department. A powerful diagnostic tool, the typical X-ray is a straightforward procedure, usually taking only seconds to complete. Even so, you'll be billed for the services of the radiology technician, use of the X-ray or CT scanner, any labs or films, as well as the expertise of the radiologist, a medical doctor whose job it is to read the scans and assist with diagnosis.

A typical bill for this simple procedure reads as follows:

Date	Rev	Dept	Charge	Description	CPT/MOD	Qty	Amount
08/24/10	450	4230	1005	Ct Thorax W/O Contrast	71250	1	$X,XXX.XX

Your doctor may also complement any diagnostic images with a blood test or sputum culture, which means you may be visited by a phlebotomist who will draw your blood. The analysis of any blood samples takes place in a clinical laboratory, complete with a blood chemical analysis machine and a Medical Director, a specialized doctor who oversees the lab. On your bill, these services may look like this:

Date	Rev	Dept	Charge	Description	CPT/MOD	Qty	Amount
08/24/10	301	4060	2981	Basic Metabolic Panel	80048	1	$XXX.XX
08/24/10	301	4060	2451	Compr. Metabolic Panel	80053	1	$XXX.XX
08/24/10	305	4060	2371	Complete Blood Count	85025	1	$XX.XX
08/24/10	306	4060	2781	Culture Sputum	87070	1	$XX.XX

HOSPITAL STAY

If your medical condition necessitates a bona fide hospital stay, you will be admitted and billed for a room. Under such conditions, the pharmacy will also be involved in your treatment. For pneumonia, your doctor may opt to use any number of antibiotics. Hospitals have a vast arsenal of medications at their disposal to treat pneumonia, although each one comes with a host of possible complications and side effects. The responsibility for providing the proper medication, as well as monitoring a patient's progress and ensuring that there are no negative drug interactions, falls on the shoulders of the hospital pharmacist as well as the patient's doctor. These goods and services will most likely show up in various places throughout your bill:

Date	Rev	Dept	Charge	Description	CPT/MOD	Qty	Amount
08/24/10	250	4170	3092	Ceftriaxone Injection 1Gm	J3490	1	$XXX.XX
08/24/10	250	4170	3083	Ketorolac Injection 30Mg	J3490	1	$XXX.XX
08/24/10	250	4170	9957	PIP/TAZOBAC VL 3.375Gm	J3490	1	$XX.XX
08/24/10	250	4170	5467	Azithromycin Injection 500Mg	Z7610	1	$XX.XX

Once admitted to the hospital, your bill will continue to grow. You have had your X-ray, your blood has been tested, and the proper medications have been prescribed. At this point, the hospital may offer you a nebulizer, a device used to administer medication directly into your lungs in the form of a mist.

Administering treatment through use of the nebulizer increases potency and speeds the body's reaction to the medicine, while reducing side effects. On a hospital bill, it looks like this:

Date	Rev	Dept	Charge	Description	Qty	Amount
08/24/10	270	4180	1081	Nebulizer Hh Inst	1	$XXX.XX

CHAPTER 14: UNDERSTANDING YOUR HOSPITAL BILL

Still, understanding the way in which your bill is tallied is only half the battle. Unfamiliar medications, codes, hospital abbreviations, and procedures can make it nearly impossible to question a given amount or line item. Like any other kind of bill, hospital bills are also subject to computer and human error. A single incorrect procedure code, entered either by the hospital or the insurance company, could significantly alter your bill, and can be difficult to find, let alone correct.

Fortunately, patients have some special protections should they need to contest a hospital bill. As a safeguard, federal regulations entitle all patients to the documentation they need to identify a billing discrepancy, including an itemized copy of the bill, a copy of the patient's medical chart, and a copy of the pharmacy ledger showing the exact medications administered during the patient's stay. By comparing documents such as these, a sharp eye may discover errors.

When you are sick in the hospital, you want nothing more than to get better. When you're home and better, all you want is a hospital bill you can understand. You may be somewhat befuddled by all the information packed into each line, but as confusing as it may initially appear, your bill is nothing more than an objective accounting of the events and ingredients that made up your hospital stay.

SMART TIP: If you are uninsured and owe more than you can afford, ask your hospital whether you are eligible for financial aid. Uninsured patients sometimes are required to pay more than the insured, so most hospitals offer special payment programs for those suffering from financial hardships.

WHO PAYS THE BILL?

If you have health insurance, the good news is that some—or most—of your hospital bill may be paid by your insurer. Here are some common payors:

Medicare, a health insurance program administered by the federal government, provides coverage to individuals aged 65 and over, or who meet other special criteria.

Medicaid is a federal and state program for eligible individuals and families with low incomes and/or resources. California's version of Medicaid is called Medi-Cal.

Most Americans receive health coverage through the **private insurance** market, usually through their jobs. The two main types of private insurance are PPOs and HMOs.

A **PPO**, or preferred provider organization, is made up of medical doctors, hospitals, and other providers who have contracted with an insurer to provide health care at reduced rates.

An **HMO**, or health maintenance organization, is an insurance entity that provides health care coverage through hospitals, doctors, and other providers who contract with the HMO.

Workers' Compensation is medical insurance for employees injured in the course of employment. In exchange, employees give up the right to sue the employer for the accident.

COBRA provides certain former employees, retirees, spouses and dependent children the right to temporary continuation of health coverage when coverage is lost due to certain specific events (such as termination from a job).

Chapter 15: Leaving the Hospital

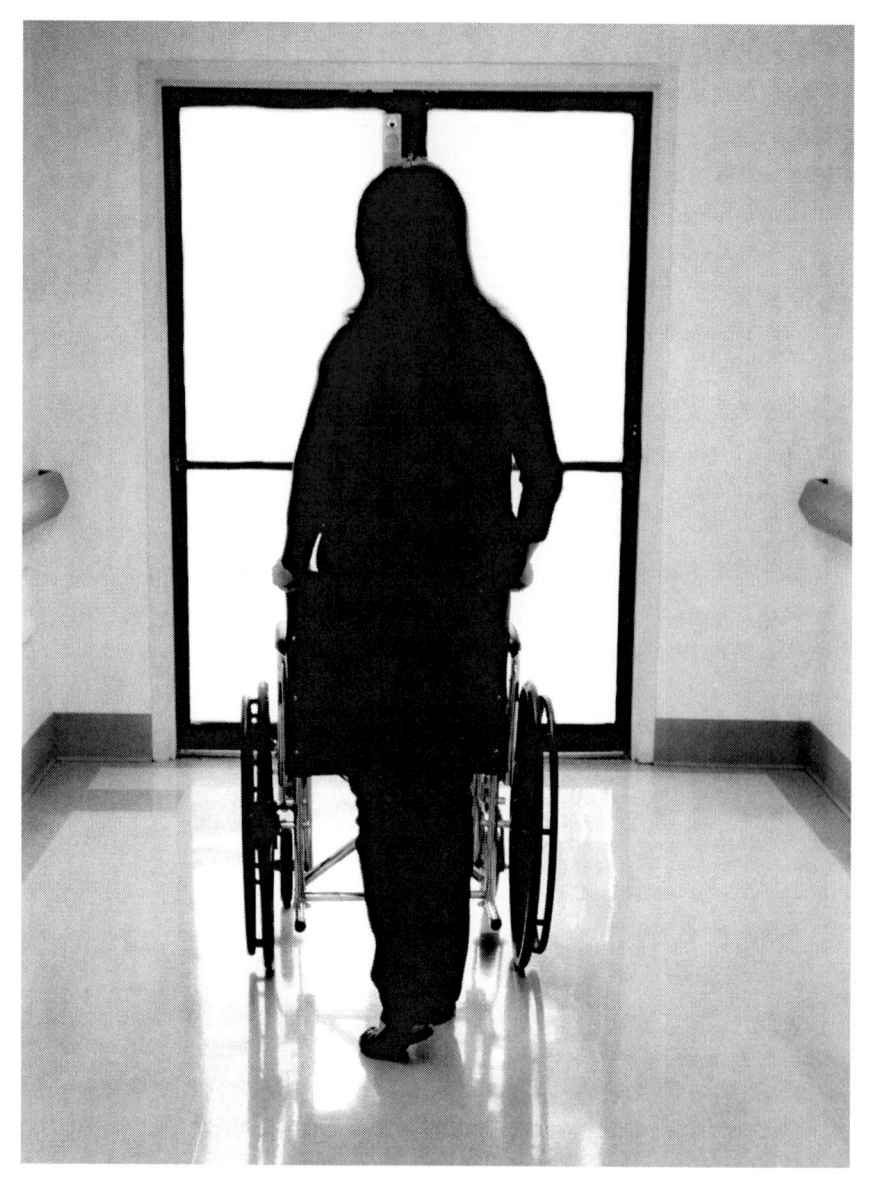

"Getting out of the hospital is a lot like resigning from a book club. You're not out until the computer says you're out."

– Erma Bombeck, American author and columnist

15

LEAVING THE HOSPITAL

TOPICS: Caregivers, Case Managers, Long-Term Care, Nursing Homes, Occupational Therapy, Physical Therapy, Subacute Care

---

One of the most welcome—and most important—stages of any hospital stay is the moment you're told you can leave. Your hospital "discharge," as the process is called, is a critical step to recovery because it includes a clear plan on how to continue your care outside the hospital. Believe it or not, your hospital starts preparing your discharge the moment you are admitted, so it's a good idea for patients and caregivers to participate in the process from the beginning. A carefully considered plan is in many ways as critical as the treatment one receives while under a doctor's direct supervision. Knowing what to expect goes a long way toward reducing stress and ensuring the best outcome for everyone involved, whether the patient is headed to his or her own home, someone else's home, or to a rehabilitation center or nursing home.

Chapter 15: Leaving the Hospital

Planning for Discharge

Discharge planning, which is offered at nearly every U.S. hospital, is a service designed to assist patients in arranging for the proper scope of care once they leave. Discharge planners can be administrators, social workers, doctors, or nurse case managers and are responsible for assisting patients and their families throughout this often complicated transition. They may work closely with families to explain a patient's needs, offer direction on continued care, and help identify the most appropriate facility to suit the situation.

Discharge planning can begin as soon as a patient is admitted to the hospital. Addressing such questions as where the patient will stay once discharged, who will oversee continued medical care, and who will assist with daily tasks helps to make the transition from the hospital to a new environment easier. Depending on the patient's condition, a good discharge plan may be as simple as taking a few days off from work to help the patient get acclimated at home or as complex as researching health care facilities and coordinating assistance among family members. Early planning enables designated caregivers to understand their upcoming roles in the days to follow the patient's discharge. It also allows them to rearrange their schedules and living quarters, if necessary. Generally, discharge is a five-stage process:

> **Stage One:** The patient's mental and physical conditions are evaluated. The attending doctor and nursing staff usually do this and may compare the patient's current state with his or her state before hospitalization. Important areas of concern may include whether the patient can safely return to his or her original living situation and whether there has been a change in the patient's ability to adequately care for him or herself. Only a doctor can order a discharge from a hospital, but a discharge planner will most likely handle the details.

Stage Two: The discharge planner explains the doctor's evaluation to the patient and any available caregivers, highlighting any special needs the patient may have outside the walls of the hospital. At this point, the focus is on future care, including whether to transfer the patient to his or her own home, that of a family member, a nursing home, or rehabilitation facility. This may include the need to decide who will be in charge of primary care, what long- and short-term services the patient will need, whether a full recovery is to be expected, and whether any new issues such as mental health or memory concerns should be addressed.

Stage Three: Once the next move has been determined, the discharge planner will begin to personalize the patient's plan. At this point, issues for discussion may include whether specific caregiver training is necessary, if third party care will be needed, and if any extra equipment (such as wheelchairs or breathing assistance devices) will be necessary. All information pertaining to the patient's medications, daily hygiene rituals, feeding concerns, and dietary needs should also be addressed in detail.

Stage Four: The discharge planner may now recommend third-party facilities or home care services that are available to suit the patient's needs, taking into consideration geographic, religious, language, and/or cultural issues that might affect quality of care. Often the discharge planner will act as liaison between the family and the chosen facility, and the planner may also provide information on government assistance programs for which the patient qualifies, as well as what the family can expect from the patient's insurance company. If necessary, issues of transportation to a new setting may also be addressed.

Chapter 15: Leaving the Hospital

Stage Five: This final phase is designed to ensure that the appointed caregiver has all the information necessary to carry out the task at hand. The discharge planner may provide a summary of the hospital stay complete with a list of results from tests and surgeries, a list of tests and/or results still pending, and a copy of the patient's discharge instructions. A list of all medications with proper dosage information should also be included as well as a 24-hour hospital access number in case of questions. Potential warning signs should be explained in detail in the event that a patient's condition should worsen. The patient should schedule a follow-up appointment either at the hospital or with a primary physician within the first two weeks of discharge to carefully monitor the patient's progress. Follow-up appointments can greatly reduce the need to readmit a patient later.

When Extra Help Is Needed

Midway through the discharge process it becomes necessary for the patient, doctor, and designated caregiver to decide upon the patient's next step. While many patients leave their primary care facility needing little more than bed rest, others face complex issues that require a more hands-on approach to rehabilitation.

Rest assured, there is a medical center to fit nearly any need the patient might have. Though this step is often stressful for the patient and caregiver, it is important to remember that help is available. When choosing a facility to assume responsibility for continued care, several factors come into play. Availability and quality of service, amount of service covered by insurance, ease of access for the caregiver, and reputation all play a crucial role in making this often difficult decision.

For those whose long-term outlooks require the services of third-party specialists, several continued care options exist:

Subacute Care Facilities
Subacute care facilities serve the needs of patients with complex medical issues who require more specialized and intensive services than those commonly found in nursing homes. This option provides highly skilled, compassionate medical, nursing, and rehabilitative care in a non-acute setting, including inhalation and respiratory assistance, wound care, and intubation management and services. Because patients at the subacute level do not require such complex diagnostic services or routine invasive procedures as those provided in acute care settings, quality outcomes can be achieved at a fraction of the cost of a fully equipped medical/surgical unit.

Subacute facilities take a goal-oriented approach to rehabilitation, often focusing on the treatment of one or more medical issues facing the patient as a result of traumatic injury, acute illness, or the exacerbation of disease-related symptoms.

Patients appropriate for subacute care include those with:

- Traumatic brain injury
- Spinal cord injury
- CVA (cerebral vascular accident)
- Neuro-muscular diseases, including multiple sclerosis, muscular dystrophy, ALS (amyotrophic lateral sclerosis or Lou Gehrig's disease), and Guillain-Barré syndrome
- Multiple organ failure
- Pulmonary rehabilitation services, including ventilator patients and tracheostomies
- Total parenteral nutrition (tube feedings)

Long-Term Care Facilities
This type of center may provide rehabilitative, restorative, and/or ongoing skilled nursing care to patients or residents in need of assistance with daily living. Depending upon the services

Chapter 15: Leaving the Hospital

offered, long-term facilities can be either acute or subacute, and include nursing homes, rehabilitation facilities, inpatient behavioral health facilities, and long-term chronic care hospitals. Those who are given long-term acute care are usually expected to make enough of a recovery to be transferred in time to a subacute counterpart, such as a nursing home.

Nursing Homes
A nursing home, convalescent home, or skilled nursing unit (SNU) provides care for patients who require constant, significant help with daily living. Residents include the elderly and younger adults with physical or mental disabilities. Patients in a skilled nursing facility may also receive physical, occupational, and other rehabilitative therapies following an accident or illness.

Things to Consider
When you or a loved one are recommended to a third-party medical facility for long- or short-term care, there are many factors to review. Because your time to decide may be limited, it is a good idea to consider the following when making your selection:

- Why was this type of facility chosen?
- What specific medical needs does this facility address?
- Is this facility capable of meeting all the patient's needs, or will additional assistance be necessary?
- How close and convenient is this facility for the primary caregivers and family?
- Is it clean, quiet, and comfortable?
- What is its reputation, both online and according to related government agencies?
- Is the structure safe and in accordance with appropriate health and safety code standards?
- Does this facility address any cultural or language related

issues the patient might have?
- Does this facility offer the services of a social worker or other liaison to interact with family and friends of the patient?

Rehabilitation Services

Whether performed during a hospital stay or after discharge, patients may require rehabilitation to maximize recovery. Based upon the illness or injury that necessitated hospitalization, rehabilitation services may vary. Here are a few examples:

Physical Therapy

This is a very broad category that involves strengthening and coordination work designed to overcome any physical limitations experienced following surgery. Some specific areas of physical therapy that may follow hospitalization include:

Cardiopulmonary rehabilitation physical therapists treat patients after cardiac or pulmonary surgery. This type of physical therapy helps increase endurance and functional independence.

Geriatric physical therapy focuses on issues specific to older patients, although this may include just about any adult patient. Certain conditions that may necessitate this particular therapy include arthritis, osteoporosis, cancer, Alzheimer's disease, hip and joint replacement, and incontinence.

Neurological physical therapy treats patients wrestling with a neurological disorder or disease, including Alzheimer's disease, ALS, cerebral palsy, multiple sclerosis, Parkinson's disease, and complications from a stroke or brain injury.

Orthopedic physical therapy focuses on musculoskeletal injuries following orthopedic surgery. The expertise of these

Chapter 15: Leaving the Hospital

physical therapists covers fractures, acute sports injuries, arthritis, sprains, strains, back and neck pain, spinal conditions, and amputations.

Speech/Swallow Therapy
If disease or injury has caused a weakness or deficit within the mouth or throat, it may present as a lack of clear speech or the inability to swallow properly. Speech therapists specialize in oral issues and provide exercises and treatment procedures that will strengthen the specific muscles that are slow or weak. This type of rehabilitation may be useful for patients with:

- Traumatic brain injury
- Stroke
- Alzheimer's disease and dementia
- Cranial nerve damage
- Progressive neurological conditions (Parkinson's, ALS, etc.)
- Autism spectrum disorders, including Asperger's syndrome
- Injuries due to complications at birth
- Feeding and swallowing difficulties
- Cerebral palsy

Occupational Therapy
Occupational therapy focuses on the patient's specific needs relating to work or performing tasks of daily living such as grooming and household care. In a hospital setting, this particular type of rehabilitation may be useful for patients with a serious medical condition due to a traumatic event such as brain or spinal cord injury. Once hospitalization has addressed and stabilized the underlying issues, occupational therapy plays an important role in facilitating early mobilization, restoring function, and preventing further decline. The goal of occupational therapy is to remove any barriers that might limit a patient's ability to live independently. At the

same time, occupational therapy attempts to provide patients with the physical functionality necessary to enjoy the same quality of leisure and community pursuits to which they were accustomed prior to the onset of disease or injury.

When the Burden of Care Falls on the Family

Patients often find themselves in a situation where their needs are not severe enough to require a third-party service, but they cannot fully care for themselves in a home setting. In such instances, a patient's family or friends may be called upon to assist during the rehabilitation process.

When a loved one returns home to recuperate, his or her needs are often diverse, and the caregiver's job can be complicated. Following are the essential elements of primary care during recovery:

Health and Hygiene: The caregiver may assume such tasks as bathing and dressing the patient, as well as assisting the patient with going to the bathroom, grooming, and eating.

Household Chores: While convalescing, the patient will most likely need help with cooking his or her food, cleaning the living quarters, and washing articles of clothing, as well as shopping for supplies and medications. The caregiver may also play the part of chauffeur when the time comes for follow-up appointments or treatments.

Medical Services: The primary caregiver will likely need to provide a certain amount of medical assistance, helping the patient with everything from wound care and bandaging to the administering of medications, including the possibility of giving injections. This can also involve performing various types of physical therapy and providing assistance in using related medical devices such as nebulizers, wheelchairs,

Chapter 15: Leaving the Hospital

monitors, or breathing apparati.

Companionship: The primary caregiver is often the one who communicates most effectively with the patient during the recovery process. The emotional aspect of rehabilitation may be directly linked to a patient's physical progress, and positive daily conversations help to reassure the patient that he or she is not facing these challenges alone.

Should the patient be suffering from Alzheimer's or have other cognitive issues, the role of a primary caregiver may take on even larger proportions. In these situations, it is important that the primary caregiver establish a relationship with the patient's doctor and support staff so that he or she can act as mediator throughout the hospital stay and be fully prepared to address the patient's needs after discharge.

If you find yourself in the position of family caregiver, know what to expect as you take on these new responsibilities. Providing post-hospitalization support for anyone can be a time-consuming, high-pressure task, and it is essential for caregivers to keep themselves both physically and mentally fit. Those who find themselves in this position should be mindful not to ignore their own needs and obligations in their effort to assist a loved one.

While the discharge process marks the conclusion of the patient's hospital stay, it is often just the first step on a long road to recovery. The challenges facing newly discharged patients and caregivers can be a complex mix of mental, physical, emotional, and financial hurdles, and the information above contains only a sampling of the issues to be faced along the way. By planning for discharge at an early stage, caregivers will be able to assemble a thoughtful strategy and prepare for added responsibilities, while the needs of the patient will continue to be met during the transition and moving forward.

A DISCHARGE CHECKLIST FOR PRIMARY CAREGIVERS

As the designated point person for your loved one's care outside the hospital, the more you know about the discharge process, the easier it will be for you to navigate the ocean of responsibilities which you will soon be forced to address.

Keep a journal throughout your loved one's hospital stay. Include the names and phone numbers of all staff members involved in treatment.

Ask questions when you do not understand any of your responsibilities, and take notes on the answers.

Ask for a written list of all tests, procedures, and treatments, as well as any pending results and follow-up treatments or procedures that need to be scheduled.

Request a functional evaluation of your loved one's current abilities, including an estimate as to how long it may take to show improvement. This step is essential in determining how much help the patient will need when brought home.

Decide where your loved one will stay after being discharged. Determine who will help with continued care, including medical as well as daily tasks.

Schedule a review of all medications **prior** to leaving the hospital. This includes dosage and frequency, and whether the medication is to be administered with or without food.

Request a 24-hour service number for questions dealing with medications, dosage, interactions, and complications.

CHAPTER 15: LEAVING THE HOSPITAL

Schedule a follow-up appointment within the next two weeks so that the patient can be evaluated by a doctor and assessed for any changes brought about by transition.

Make a list of supplies that may be needed during convalescence, including bandages, ointments, diapers, syringes, disposable gloves, and any other necessities.

Request a list of any potential problems, including symptoms, side effects, or areas of concern. This includes issues of medication and wound treatment as well.

Inquire about nutritional diet related issues that may need to be closely watched.

Ask for training if your loved one requires medical treatment with which you are unfamiliar. This is especially important if additional equipment such as an oxygen machine or heart monitor is needed. Request a direct telephone number for issues or concerns relating to any equipment or procedure.

Call the health insurer, if any, and ask about the services and coverage available to the patient. Understand exactly what the patient will and won't be billed for.

Request a list of support groups that specialize in the patient's illness, as well as any organizations that provide counseling for the needs of caregivers.

Thank the discharge planner and any support staff for their roles in assisting your loved one throughout his or her stay.

HOSPITAL STAY

CHAPTER 16: WELCOME TO THE MORGUE

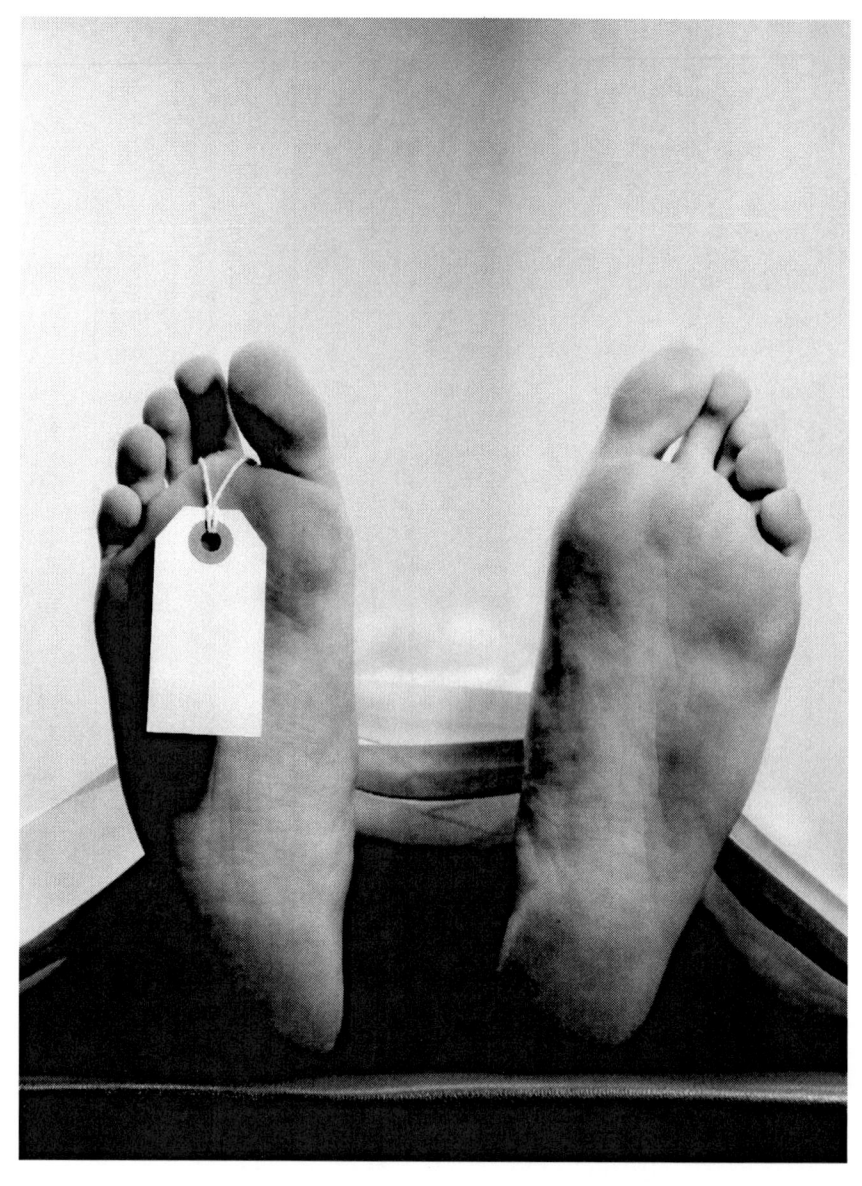

"All interest in disease and death is only another expression of interest in life."
— Thomas Mann, German author, *The Magic Mountain*

16

WELCOME TO THE MORGUE

TOPICS: Autopsy, Body Bags, Forensic Pathologist, Medical Examiner, Organ Harvesting and Donation, Religious Concerns

No one wants to be told that the morgue awaits them or their loved ones, but you might be curious about what goes on there. Morgues, found in some hospitals, funeral homes, and state or county facilities, play an essential role in the cultural, medical, and legal processing of the deceased. A Latin proverb found on the walls of many morgues—*Taceant colloquia. Effugiat risus. Hic locus est ubi mors gaudet succurrere vitae*, or "Let conversation cease. Let laughter flee. This is a place where death delights to help the living"—suggests that death is not the end, at least from a medical standpoint. Procedures performed in the morgue can help families understand the cause of a loved one's death, further educational research, and sometimes aid in legal proceedings.

Fortunately, what you've seen on TV about the morgue probably won't happen to you or anyone you know. Family members

Chapter 16: Welcome to the Morgue

usually visit the morgue only if the deceased needs to be identified. If someone passes away in a hospital, nursing facility, or residence, identification is seldom an issue, and the body of the deceased is often taken directly to the chosen funeral home's morgue and prepared for burial or cremation according to the family's wishes. Here's a quick look at what can happen behind the scenes.

What's in the Morgue
Modern morgues are clean, isolated areas where bodies can be preserved until identified, claimed, prepared for burial, or if necessary, autopsied to confirm cause of death. Since fifth century Paris, morgues have functioned as the penultimate resting place of the dead, and even today they play a valuable safety role in preventing the spread of communicable diseases.

In a typical morgue, tables, refrigerated storage units, flooring, and wall coverings are often made of stainless steel, because it is easy to clean and keep cool, which aid in slowing or preventing decomposition. In addition, morgues have what is known as a "cold chamber," a specially designed, temperature regulated storage facility where bodies are kept. There are two types of cold chamber: ones that are set to a temperature that slows but does not deter decomposition and ones that are chilled enough to freeze the body and prevent decomposition. The latter are most often used when identification is an issue.

Today's morgues don't have bells and attendants on patrol, as some did in the eighteenth and nineteenth centuries, but they are staffed to cover a variety of duties. (See "Who's Who" sidebar.) It is also reassuring to know that hospitals go to great lengths to ensure patients are not erroneously brought to the morgue while still alive. This was a concern in past centuries when laypeople more often than medical doctors were the ones to determine whether a person had died.

Body Bags

Body bags, also known as "cadaver pouches," are found in every morgue. These bags are normally made of thick white or black plastic, so that evidence that falls from the body can be easily seen and recovered after transport. They are equipped with a zipper and are non-porous to prevent the normal leakage of bodily fluids after death.

Pertinent information about the deceased may be written in marker on the body bag, either for investigative use or for the purposes of identification at the morgue before entering storage. Body bags are not reusable and are usually incinerated after the corpse has been buried or cremated. In times of war it is not unusual for body bags to function as temporary coffins for fallen soldiers until more permanent accommodations can be made.

Why Autopsies Are Performed

An autopsy, also known as a "post-mortem" or "necropsy," may be performed for legal or medical purposes either in a hospital morgue (by a pathologist or forensic examiner) or a county morgue (by the medical examiner or coroner). During an autopsy, the body is dissected, enabling a detailed internal and external examination. X-rays and lab tests are often run to detect disease, infection, drugs, or poisons. A careful examination can shed light on the individual's health prior to death, and may help determine whether a hospital patient's medical treatment and diagnosis were appropriate. For some families, this is an important step in coming to terms with a loved one's death.

The laws for ordering and performing autopsies vary greatly from state to state. Most deaths do not require an autopsy, but one is nearly always conducted if foul play is suspected (e.g., a homicide). In most states, an autopsy may also be ordered if the general public health is believed to be at risk or if an attending physician does not wish to sign the death certificate—which can

Chapter 16: Welcome to the Morgue

happen if the patient died unattended and death was sudden and unaccountable (in other words, the cause is in question). In addition, an autopsy can be performed for any reason at the family's request, often free of cost.

There are two main categories of autopsy:

> **Forensic autopsies** are legal in nature and are used to determine the cause and manner of death, as well as to identify the body if necessary. They are most often performed in the event of a suspicious, sudden or violent death, or after certain surgical procedures to rule out medical error. Under U.S. law, every death must be classified under one of the following categories (frequently determined via autopsy):
>
> - Natural
> - Accident
> - Homicide
> - Suicide
> - Undetermined
>
> **Clinical autopsies** are performed for research or educational purposes. They can also be used to study matters that may affect public health, like an outbreak of any contagious disease. They can determine whether a doctor's diagnosis was correct and identify which outstanding issues contributed to a patient's death. Clinical autopsies are designed to help doctors learn from their experiences in the field, as well as to ensure consistency of care in hospitals.

Religious Concerns

If you have religious concerns about an autopsy, it's important to make your wishes known to the hospital. Once made aware, a pathologist or mortician will consider the religion of any deceased person who ends up on his or her table. Most cultures follow the principle that the deceased should be buried within 72 hours to prevent decay and ensure a lifelike image, while others opt to have the body preserved for several months. Orthodox Jews often insist on having a rabbi present during an autopsy, while many Muslims do not condone the practice at all. This can make the pathologist's job a difficult one, especially if required to perform an autopsy by law and in a timely fashion. A state can legally perform an autopsy over the objections of the family of the deceased if criminal activity is suspected or if it is thought to be of public necessity.

Organ Harvesting and Donation
Another procedure that is associated with the morgue, although it doesn't technically take place there, is organ harvesting. Organ harvesting is the removal and preservation of human organs or tissue from the bodies of the recently deceased to be used in surgical transplants on the living. It is performed in an operating room by licensed surgeons, and unlike many autopsies, does not disfigure the body, allowing the family of the deceased to hold an open casket funeral or memorial service.

Although mired in ethical debate and heavily regulated, organ harvesting and donation in the United States has become an accepted medical practice. According to the American Transplant Foundation, a nonprofit organization dedicated to increasing organ and tissue donation, more than 106,000 people in the U.S. were waiting for a life-saving organ transplant in 2010.

The process of harvesting organs is a complex and delicate issue. Despite urban myths you may have heard, organ donation is only considered once it is clear that a patient has no hope of survival. The prospect of gathering viable organs from a severely injured patient plays no role whatsoever in a doctor's overall diagnosis or treatment. A patient may elect in advance to be an organ

CHAPTER 16: WELCOME TO THE MORGUE

and tissue donor, or family members (including individuals holding power of attorney for a brain dead patient) can give consent to donate post mortem. There is never a risk that organ harvesting will take place without proper consent.

Aside from certain organs such as kidneys that can come from live donors, brain death is a prerequisite for donating an organ. This means the patient shows no brain activity, though the heart continues to beat and breathing is regulated by a ventilator. While the patient has no hope for recovery and is considered legally dead, the organs are still receiving the necessary life forces to keep them in healthy working order and viable as transplants. Once a suitable candidate is found, the first step is confirmation of death. A series of tests are conducted as quickly as possible to ascertain that this is the case.

When a more precise determination is needed, a number of diagnostic lab tests can be performed such as brain scans, computerized mapping of the brain (tomography), testing or taking pictures of brain blood flow, or measuring the electrical activity of the brain or brain stem. For even more certainty, these tests may be repeated hours later.

Once all legal issues have been addressed, a transplant team initiates the organ harvesting process. To keep the organs healthy and fully functional, the team must work quickly. Some tissues have a shelf life and may be stored in medical facilities until they are needed, but most often healthy organs and tissues have a better chance of survival if the transplant occurs quickly. A heart, for example, should be used within six hours. Operations to remove a kidney, the most frequently transplanted of all organs, and transplant it into the recipient are done within the same day—often in the same hospital. As soon as the medical team has received the necessary approval, a transplant coordination agency, such as the United Network for Organ Sharing, is notified and the patient's organs are quickly yet respectfully removed and prepared for

> **BRAIN DEATH IS DETERMINED BY THREE FACTORS:**
> - Coma and total lack of response to painful stimuli.
> - Total lack of brain stem reflexes—determined by the size of pupils and reaction to light, response of eyeballs with rapid movement of head and corneal stimulation, response to cold water in ear, reaction to stimulation of cough and gag reflexes.
> - Total lack of spontaneous respiration—determined by providing patient with pure oxygen and watching for signs of respiration.

transit. In the event that a donated organ or tissue specimen is not usable, it is properly disposed of. Upon completion, the body is sewn up and restored so that the family may conduct a viewing and prepare their loved one for burial.

Old Fears

The fear of being mistaken for dead has been around since time immemorial. It reached a fevered pitch in the eighteenth and nineteenth centuries, when rumors about premature burial circulated widely. In response to public concern, some German towns constructed buildings called waiting mortuaries to store the recently deceased until decomposition began, as this was a more reliable sign of death than lack of pulse or breathing. Waiting mortuary attendants patrolled the premises, looking for signs of life among the new arrivals. In some cases, strings were affixed from the insides of coffins to an external bell in the hope that, were a misdiagnosis made, the attendant would be alerted and the unfortunate soul saved from being buried alive. After more than 200 years of use, however, there were no credible records of even a single life being saved, so waiting mortuaries eventually fell out of favor and were discontinued.

Chapter 16: Welcome to the Morgue

WHO'S WHO IN THE MORGUE

Most morgues in the U.S. are headed by a **forensic pathologist**, a medical or osteopathic doctor trained in forensic science and specializing in diagnosing diseases and issues through inspection of the body's organs and tissues. As anyone who has seen a television crime drama like *CSI* will tell you, forensic pathologists have a wide range of responsibilities, from assisting in death-scene investigations and performing autopsies to reviewing medical records, determining cause of death, and giving testimony in court.

In comparison, the position of **coroner** is a public office. The term is derived from the Latin *corona*, meaning "crown," as in "one in the service of the king," though today's coroners may be either elected or appointed. Modern day coroners are public officials who investigate and certify cause and manner of death in sudden and unnatural cases. The specifics of both their job and training vary greatly from state to state. Qualifications run the gamut from laymen to physicians, law enforcement officers, and accredited pathologists.

The office of the **medical examiner (ME)** was created in the late nineteenth century, as a result of mistrust of coroners who were often unqualified and corrupt. Today, both coroners and MEs are public officials whose duty is to investigate the cause and manner of suspicious, sudden and unnatural deaths within their district. MEs, however, are always appointed, whereas coroners are sometimes elected. Though they may be trained in any medical specialty, they are usually forensic pathologists.

HOSPITAL STAY

Chapter 17: Complementary And Alternative Medicine

"Nature, time, and patience are three great physicians."
— Bulgarian proverb

17

COMPLEMENTARY AND ALTERNATIVE MEDICINE

TOPICS: Acupuncture, Complementary and Alternative Medicine (CAM), History, Holistic Approach, Yoga

Given our most common associations with hospitals—high-tech equipment, sterile rooms, and modern surgical procedures—treatments such as massage, acupuncture and meditation may seem to be at the opposite end of the spectrum in the health care arena. But in actuality, an increasing number of hospitals are incorporating treatments such as these into their services.

Complementary and alternative medicine (CAM) includes yoga, acupuncture, meditation, and many more options considered outside of traditional Western medicine, and is now offered in almost 40 percent of hospitals, according to Health Forum, a subsidiary of the American Hospital Association. More than 37 percent of hospitals surveyed in 2007 offered one or more CAM therapies, up from 26.5 percent in 2005, and 8 percent in 1998. And it is in

Chapter 17: Complementary And Alternative Medicine

fact some of the most modern, top-rated hospitals that are leading the pack. All 14 of the top hospitals listed in the *U.S. News & World Report's* 2010-2011 Best Hospitals Honor Roll offer complementary and alternative services in some form, and many of them—such as Duke University Medical Center, University of California, San Francisco Medical Center, and the Mayo Clinic—are home to clinics dedicated to CAM services and clinical research.

What Is CAM?

Complementary and alternative medicine includes numerous and varied treatments that fall outside of traditional Western medicine, yet increasingly appeal to patients for general wellness options, preventative measures, or when conventional choices fall short, such as when treating chronic pain.

According to the National Center for Complementary and Alternative Medicine (NCCAM), "CAM is a group of diverse medical and health care systems, practices, and products that are not generally considered to be part of conventional medicine." Western medicine is associated with the most advanced, research-based treatments, surgeries, and drugs used by medical doctors and their colleagues. CAM treatments may date back thousands of years to ancient Greece or China and include acupuncture, herbal medicine, chiropractic treatments, hydrotherapy, art therapy, and more.

Previously referred to as "alternative medicine," the phrase "complementary and alternative medicine" has gained popularity to avoid the stigma of being labeled as unorthodox. "Alternative" is still used among some practitioners and patients when a treatment is providing a real substitute for Western medicine, but more and more, health care practitioners use CAM therapies concurrently with Western practices. Clinical studies suggest that acupuncture, for example, can help relieve nausea from chemotherapy, and the Memorial Sloan-Kettering Cancer Center in New York uses therapies such as acupuncture, hypnotherapy and meditation in addition to mainstream treatments to help relieve the pain and stress often accompanying cancer treatment. The buzz word "integrative," in fact, is also growing in popularity. Making use of both scientifically-validated CAM and conventional therapies to treat a patient is referred to as "integrative medicine."

A Decade Of Integration
This gradual transformation from the alternative to complementary mindset has been born on CAM's increased integration over the past 10 years with traditional medical establishments such as clinics and hospitals. The American Hospital Association's Health Forum survey suggests patients are driving the transition: 84 percent of hospitals cited patient demand as the primary reason for offering CAM services.

"Today's patients have better access to health information and are demanding more personalized care," study author Sita Ananth noted in the release. According to the U.S. government 2007 National Health Interview Survey, CAM patients spent $33.9 billion for out-of-pocket costs on products and services. While some insurance plans offer CAM coverage, this varies by region and is often limited. Most people, according to NCCAM, pay for CAM products and services themselves. Although usage is highest among women and those with higher incomes and levels of education, people of all backgrounds use complementary medicine today. About 4 out of 10 adults are using some form of CAM in the U.S., including programs designed for men and children, such as UCLA's Pediatric Pain Program.

In addition, federal research funds and private donations have prompted health groups to expand research and clinical trials. The National Institutes of Health's funding of complementary medicine was nearly $300 million in 2009 compared to $116 million in 1999. After academic centers such as the University of Arizona and Stanford University opened integrative medicine

Chapter 17: Complementary And Alternative Medicine

clinics in the 1990s, a growing number have followed suit in an effort to both offer patients services and research the effectiveness of such treatments with the scientific rigor of Western medicine.

Given the broad range in the types of complementary therapies, some hold more promise than others. And just like Western medicine, the CAM sector is not without its detractors. Some attribute success with complementary therapies to the placebo effect—a measurable or felt improvement inspired purely by a patient's belief that an inert substance they're taking is an active treatment. Others point to the body's own natural healing process.

Considering the nascent stage of CAM therapies in Western medicine, evidence-based research is critical for continued growth and integration. Western medicine is based upon medical research, where treatment results are tested, reproduced, and subject to strict safety and efficacy protocols. Many CAM treatments have yet to go through that same scrutiny.

At the same time, 67 percent of hospitals noted clinical effectiveness as their top reason for offering CAM services in the American Hospital Association's Health Forum survey. Centers like Mayo Clinic's Complementary and Integrative Medicine program are conducting numerous clinical studies to identify and share information about leading CAM therapies with physicians and their patients.

Holistic Approach

With so many therapies falling under the umbrella of complementary and alternative, there are a myriad of different ways that hospitals and medical centers might integrate CAM into their establishments. While hospitals typically focus on the more established CAM therapies like massage, relaxation training, and nutritional therapy, even these treatments typically vary.

Still, integrative centers at hospitals often share a common approach in treating the whole person—their physical, psychological, and spiritual selves—rather than just a specific symptom. The holistic approach recognizes the interconnectedness of a patient's body, emotions, attitudes, and environment. A center's doctor may look at all of these pieces in order to understand the root cause of a known ailment, to come up with a diagnosis, or to determine a treatment. In addition to body-based therapies such as acupuncture, Mayo Clinic's CAM program, for example, offers stress management, hypnosis, and resilience training to help patients develop positive coping strategies. At the Duke Integrative Medicine center, key wellness areas include environment, relationships, personal growth, and spirituality as well as nutrition and exercise.

Some of the treatments offered at centers like the above are in and of themselves holistic systems, such as traditional Chinese medicine and Ayurveda. Traditional Chinese medicine incorporates acupuncture, herbal medicine, massage, dietary, and lifestyle treatments, among others, while Ayurveda is a traditional Indian medical system whose treatments includes yoga, diet, massage, and meditation.

This does not mean that complementary treatments are always holistic—many therapies are used to treat a single symptom, and likewise Western medicine is not always reductionist. A patient's background and habits will often be considered for diagnosis and treatments. But the holistic approach that characterizes many CAM centers is a lure for patients seeking "whole-person" care.

Common CAM Therapies

Some of the most common conditions treated with complementary and alternative medicine at hospitals include chronic pain, cancer, preparation for surgery and/or recovery, women's health, and anxiety and depression. However, patients seek CAM treatments for any number of conditions, including chronic fatigue, sinusitus, addiction, gastrointestinal conditions, sports injuries, or as a preventative measure.

Chapter 17: Complementary And Alternative Medicine

As with CAM-treated ailments, CAM treatments are far-ranging. They vary in cultural origin, philosophy, and history—some may be thousands of years old, others only decades. They may include anything from homeopathy to pet therapy, an increasingly popular inpatient service, sometimes referred to as Animal Assisted Therapy (AAT). The following are some of the most common treatments found at hospitals and integrative centers in the U.S.

Acupuncture involves the insertion of fine needles at specific body points and is believed to rebalance energy flows (qi) along pathways (meridians) to alleviate pain or treat conditions such as insomnia and high blood pressure. Acupressure is similar but involves the application of pressure as opposed to needles.

Biofeedback uses electronic devices to enhance patients' control over body functions, such as heart rate and muscle tension, to treat conditions like anxiety, insomnia, and migraines.

Chiropractic body manipulation is based on diagnosis and manipulation of joints used primarily to treat spine, joint, and muscle problems.

Creative therapy uses creative expression—music, art, dance, and color, for example—for therapeutic purposes, such as dealing with difficult emotions or stress.

Herbal medicine, rooted in Western and Eastern traditions such as Ayurvedic, Chinese, Japanese, and Tibetan, uses medicinal fruits and plants to promote healing.

Hypnotherapy uses therapeutic suggestions to the mind in a relaxed state to encourage healing of physical or emotional problems, such as phobias, addictions and anxiety. Hypnotherapy is an example of "mind-body therapy" which, according to NCCAM, uses techniques "to enhance the mind's capacity to affect bodily function and symptoms."

Massage is the application of pressure to induce relaxation and is believed to include a range of health benefits such as improved immunity and circulation.

Meditation, a form of mind-body therapy, uses a variety of techniques—such as focusing on a word or image—to clear the mind and induce relaxation. Other techniques include deep breathing, progressive relaxation and guided imagery.

Nutrition therapy uses food and supplements tailored to an individual to treat ailments or as a preventative measure.

Qi gong, Ta'i chi, and yoga: A part of the traditional Chinese medicine system, Qi gong is a system of static and moving exercises that involves breathing and meditation. Also of Chinese origin, Ta'i chi involves slow, flowing movements and breathing techniques. Yoga, a part of Ayurvedic medicine, is a system of exercises, breathing, and meditation techniques.

Reiki, or "universal life energy," involves transferring healing energy through hand positions to treat and rebalance the body.

Whole medical systems, according to NCCAM, are complete systems of theory and practice, often having evolved over centuries, such as Ayurveda and traditional Chinese medicine. Naturopathy is another conventional whole medical system with a focus on utilizing natural treatments such as herbal medicine, acupuncture, and nutrition to encourage the body to heal itself and return to a state of balance.

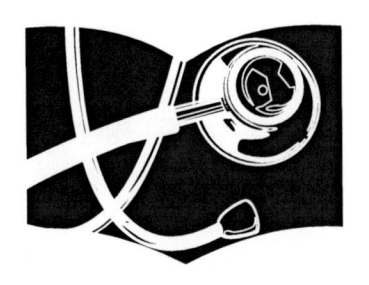

Join the conversation at HospitalStay.com

Want to learn more?
Additional resources and news are available on our website at www.HospitalStay.com. You can also follow us on Twitter @hospitalstay.

Want to buy this book?
You can purchase copies of this book online at: www.HospitalStay.com